Power Systems

J. F. Gieras

Advancements in Electric Machines

Jacek F. Gieras

Advancements in Electric Machines

With 235 Figures

Springer

Jacek F. Gieras, FIEEE
UTC Hamilton Sundstrand Fellow
Applied Research, Rockford, Illinois, USA
Full Professor of Electrical Eng.
University of Technology and Life Sciences
Bydgoszcz, Poland
jgieras@ieee.org

ISBN: 978-1-4020-9006-6   e-ISBN: 978-1-4020-9007-3

Power Systems ISSN: 1612-1287

Library of Congress Control Number: 2008932939

This work is subject to copyright. All rights are reserved, whether the whole or part of the material is concerned, specifically the rights of translation, reprinting, reuse of illustrations, recitation, broadcasting, reproduction on microfilm or in any other way, and storage in data banks. Duplication of this publication or parts thereof is permitted only under the provisions of the German Copyright Law of September 9, 1965, in its current version, and permission for use must always be obtained from Springer. Violations are liable for prosecution under the German Copyright Law.

The use of general descriptive names, registered names, trademarks, etc. in this publication does not imply, even in the absence of a specific statement, that such names are exempt from the relevant protective laws and regulations and therefore free for general use.

Cover design: deblik, Bauer, Thomas

Printed on acid-free paper

9 8 7 6 5 4 3 2 1

springer.com

Preface

Traditionally, electrical machines are classified into d.c. commutator (brushed) machines, induction (asynchronous) machines and synchronous machines. These three types of electrical machines are still regarded in many academic curricula as fundamental types, despite that d.c. brushed machines (except small machines) have been gradually abandoned and PM brushless machines (PMBM) and switched reluctance machines (SRM) have been in mass production and use for at least two decades.

Recently, new topologies of high torque density motors, high speed motors, integrated motor drives and special motors have been developed. Progress in electric machines technology is stimulated by new materials, new areas of applications, impact of power electronics, need for energy saving and new technological challenges. The development of electric machines in the next few years will mostly be stimulated by computer hardware, residential and public applications and transportation systems (land, sea and air).

At many Universities teaching and research strategy oriented towards electrical machinery is not up to date and has not been changed in some countries almost since the end of the WWII. In spite of many excellent academic research achievements, the academia–industry collaboration and technology transfer are underestimated or, quite often, neglected. Underestimation of the role of industry, unfamiliarity with new trends and restraint from technology transfer results, with time, in lack of external financial support and drastic decline in the number of students interested in Power Electrical Engineering. It is not true that today's students are less interested in heavy current electrical engineering, but it is true that many educators have little to offer them.

Universities are no longer leaders in the development of modern electrical machinery. A quick study of recent advancements in electrical machinery, patent database, web sites of funding agencies and the list of references given at the end of this book shows that most inventions related to electric machinery have been done by practising engineers. Technological breakthroughs are born in small companies established *ad hoc* for the development of a specific product and R&D centers of large corporations.

This book discusses the state-of-the art of electric machines, associated components and contemporary trends of their development. Novel electric machines considered in this book have been limited to rotary electric machines. Linear electric machines and linear actuators have not been included. There are three objectives of this book:

- to promote electrical machines as the most popular machines of everyday life and increase awareness in this area;
- to stimulate innovations in electrical machines and electromechanical drives;
- help educators to revitalize Power Engineering curricula and change research strategy towards what industry and mankind needs.

Looking at the map of the world, research and development activity in electrical machines since the end of 19th Century has been located in Central Europe, the U.K., the United States and the Soviet Union. After the WWII, Japan joined this group of leaders and recently South Korea. Development of modern electric machines and electromechanical drives require large injection of funds, unavailable in most countries, so that the role of the European Union and Russia has been gradually diminishing. Without any doubt, the United States is now the world leader in the development of new types of electric machines, their nonstandard topologies and application of novel materials to their construction.

I want to thank the people that were helpful in making this book possible. Mr William B. Kroll, the Director of Systems and Engineering at Hamilton Sundstrand has encouraged me to write this book and mentored me in aerospace business. Professor Miroslaw Dabrowski of the Technical University of Poznan, Poland has read Chapter 1. Professor Janusz Turowski of the Technical University of Lodz, Poland has read Chapter 10. Dr Roger Wang of the University of Stellenbosch, South Africa has read Chapter 3. Dr Robert Chin of ABB Sweden has read Chapter 9. My daughter Izabella Anna Gieras, the Director of Technology Management at Beaumont Hospital, Royal Oak, MI, U.S.A. has read Chapter 6. Mr William B. Kroll, Dr Stephen R. Jones and Ms Colleen Carroll of Hamilton Sundstrand have read the whole book. They have all made many useful suggestions.

Many thanks to Ms Anneke Pott and Ms Nathalie Jacobs of Springer for very efficient collaboration and assistance in preparation of this book.

Last, but not least, I would like to thank my wife Janina, and our children Izabella Anna, Karolina Maria and Michael Benjamin, for their love, patience, sacrifice, and understanding. I feel sorry that I have never had enough time for them.

Any comments, criticism, and suggestions for improvements are most welcome and may be sent to the author at jgieras@ieee.org

Rockford, Illinois, U.S.A., March 2008 *Jacek F. Gieras*

Contents

1. **Introduction** 1
   1.1 Why electric machines continue to naturally grow 1
   1.2 Status of electric motors 4
       1.2.1 A.c. motors 4
       1.2.2 Brushless PM motors 6
       1.2.3 Stepping motors 9
       1.2.4 Switched reluctance motors 10
       1.2.5 Servo motors 11
   1.3 Progress in electric machines technology 12
   1.4 Mechatronics 13
   1.5 Microelectromechanical systems 13
   1.6 Superconductivity 14
   1.7 Solid state converters 19
   1.8 Energy conservation 21
   1.9 Power quality 23
   1.10 Recyclable electric machines 24

2. **Material engineering** 27
   2.1 Laminated silicon steels 27
   2.2 High saturation ferromagnetic alloys 31
   2.3 Amorphous ferromagnetic materials 36
   2.4 Soft magnetic powder composites 36
   2.5 Permanent magnets 43
       2.5.1 Characteristics of PM materials 43
       2.5.2 Rare-earth permanent magnets 44
       2.5.3 Halbach array 47
   2.6 Wire insulation with heat activated adhesives 48
   2.7 High temperature materials 49
       2.7.1 High temperature ferromagnetic materials 49
       2.7.2 High temperature insulating materials and conductors . 49
   2.8 Superconductors 50

		2.8.1	Classification of HTS wires	50

         2.8.1   Classification of HTS wires 50
         2.8.2   HTS wires manufactured by American Superconductors 53
         2.8.3   HTS wires manufactured by SuperPower 56
         2.8.4   Bulk superconductors 58
  2.9   Nanostructured materials................................. 62
         2.9.1   Carbon nanotubes 62
         2.9.2   Soft magnetic nanocrystalline composites 65
  2.10   Magnetic shape memory materials 68

3 High power density machines 71
  3.1   Permanent magnet transverse flux motors 71
  3.2   Permanent magnet disc type motors 76
  3.3   Permanent magnet motors with concentrated non-overlapping
         coils 77
  3.4   Motors for refrigeration compressors 78
  3.5   Induction motors with cryogenic cooling system 79

4 High speed machines 81
  4.1   Requirements 81
  4.2   Microturbines 82
  4.3   Compressors 88
  4.4   Aircraft generators 89
  4.5   High speed multimegawatt generators 96
         4.5.1   Directed energy weapons 96
         4.5.2   Airborne radar 97
         4.5.3   Megawatt airborne generator cooling system 99
  4.6   Comparison of cooling techniques for high speed electric
         machines 102
  4.7   Induction machines with cage rotors 103
  4.8   Induction machines with solid rotors 109

5 Other types of novel motors 115
  5.1   Written pole motor 115
  5.2   Piezoelectric motors 118
  5.3   Bearingless motors.....................................119
  5.4   Slotless motors 121
  5.5   Coreless stator permanent magnet brushless motors..........123
         5.5.1   Disc type coreless motors........................123
         5.5.2   Cylindrical type motors with coreless stator winding ...126
  5.6   Integrated starter generator.............................129
  5.7   Integrated electromechanical drives 130
  5.8   Induction motors with copper cage rotor 131

6 Electric motors for medical and clinical applications 135
6.1 Electric motors and actuators 135
6.2 Material requirements 138
6.3 Control 139
6.4 Implanted blood pumps 140
6.5 Motorized catheters 146
6.6 Plaque excision 147
6.7 Capsule endoscopy 148
6.8 Minimally invasive surgery 151
6.9 Challenges 156

7 Generators for portable power applications 157
7.1 Miniature rotary generators 157
    7.1.1 Mini generators for soldiers at battlefields and unmanned vehicles 157
    7.1.2 Coreless stator disc type microgenerators 163
7.2 Energy harvesting devices 164

8 Superconducting electric machines 171
8.1 Low speed HTS machines 171
    8.1.1 Applications 171
    8.1.2 Requirements 173
    8.1.3 HTS synchronous motor for ship propulsion rated at 5 MW 174
    8.1.4 Test facility for 5 MW motors 177
    8.1.5 HTS motor for ship propulsion rated 36.5 MW 180
    8.1.6 Superconducting synchronous generators 183
    8.1.7 Dynamic synchronous condenser 185
    8.1.8 HTS synchronous generators developed by *Siemens* 185
    8.1.9 Japanese HTS machines 189
    8.1.10 Bulk HTS machines 196
    8.1.11 HTS synchronous generator built in Russia 198
    8.1.12 HTS d.c. homopolar generator 199
8.2 High speed HTS generators 202
    8.2.1 First prototype of high speed superconducting generators 202
    8.2.2 Homopolar generators with stationary superconducting winding 205
    8.2.3 Design of HTS rotors for synchronous generators 207
8.3 Market readiness 209

9 Naval electric machines 213
9.1 Background 213
9.2 Power train of electric ships 217
9.3 Propulsion units 218

9.3.1	Shaft propulsion 218
9.3.2	Azimuth thrusters 219
9.3.3	Pod propulsors 220
9.3.4	Integrated motor-propeller 221

9.4 Generators for naval applications 223
9.5 Electric motors for naval applications 224
    9.5.1 Large induction motors 224
    9.5.2 Large wound rotor synchronous motors 225
    9.5.3 Large PM motors 225
    9.5.4 Axial flux disc type PM brushless motors 226
    9.5.5 Transverse flux motors 228
    9.5.6 IMP motors 229
    9.5.7 Superconducting motors 234

10 Scenario for nearest future 235
10.1 Computer hardware 235
    10.1.1 Hard disc drive motors 235
    10.1.2 Cooling fan motors 237
10.2 Residential and public applications 238
    10.2.1 Residential applications 238
    10.2.2 Public life applications 241
    10.2.3 Automotive applications 242
10.3 Land, sea and air transportation 242
    10.3.1 Hybrid electric and electric vehicles 242
    10.3.2 Marine propulsion 246
    10.3.3 Electric aircraft 247
10.4 Future trends 251

Abbreviations 255

References 259

Index 271

Author's Biography

Jacek F. Gieras graduated in 1971 from the Technical University of Lodz, Poland, with distinction. He received his PhD degree in Electrical Engineering (Electrical Machines) in 1975 and Dr habil. degree (corresponding to DSc), also in Electrical Engineering, in 1980 from the University of Technology, Poznan, Poland. His research area is Electrical Machines, Drives, Electromagnetics, Power Systems, and Railway Engineering. From 1971 to 1998 he pursued his academic career at several Universities worldwide including Poland (Technical University of Poznan and Academy of Technology and Agriculture Bydgoszcz), Canada (Queen's University, Kingston, Ontario), Jordan (Jordan University of Sciences and Technology, Irbid) and South Africa (University of Cape Town). He was also a Central Japan Railway Company Visiting Professor at the University of Tokyo (Endowed Chair in Transportation Systems Engineering), Guest Professor at Chungbuk National University, Cheongju, South Korea, and Guest Professor at University of Rome *La Sapienza*, Italy. In 1987 he was promoted to the rank of Full Professor (life title given by the President of the Republic of Poland). Since 1998 he has been affiliated with United Technology Corporation, U.S.A., most recently with Hamilton Sundstrand Applied Research. In 2007 he also became Faculty Member of the University of Technology and Life Sciences in Bydgoszcz, Poland.

Prof. Gieras authored and co-authored 10 books, over 240 scientific and technical papers and 30 patents. The most important books are "Linear

Induction Motors", Oxford University Press, 1994, U.K., "Permanent Magnet Motors Technology: Design and Applications", Marcel Dekker Inc., New York, 1996, second edition 2002 (co-author M. Wing), "Linear Synchronous Motors: Transportation and Automation Systems", CRC Press LLC, Boca Raton, Florida, 1999, "Axial Flux Permanent Magnet Brushless Machines", Springer-Kluwer Academic Publishers, Boston-Dordrecht-London, 2004, second edition 2008 (co-authors R. Wang and M.J. Kamper) and "Noise of Polyphase Electric Motors", CRC Press - Taylor & Francis, 2005 (co-authors C. Wang and J.C. Lai). He is a Fellow of IEEE , U.S.A., Full Member of International Academy of Electrical Engineering, Hamilton Sundstrand Fellow (U.S.A.) and member of Steering Committees of numerous international conferences. He is cited by *Who's Who in the World*, Marquis, U.S.A., 1995–2009, *Who's Who in Science and Technology*, Marquis, U.S.A., 1996–2009, *Who's Who in Finance and Industry*, Marquis, U.S.A., 1998–2009, *Who's Who in America*, 2000–2009, and many other dictionaries of international biographies.

1
Introduction

In this chapter the state-of-the art of electric machines, associated components and contemporary trends of their development have been discussed. Recently, new topologies of high torque density motors, high speed motors, integrated motor drives and special motors have been developed. Progress in electric machines technology is stimulated amongst others by new materials, new areas of applications, impact of power electronics, need for energy saving and new technological challenges. The development of electric machines in the next few years will mostly be stimulated by computer hardware, residential and public applications and transportation systems (land, sea and air).

1.1 Why electric machines continue to naturally grow

Electric machines are the most popular machines of everyday life and their number of types increases with developments in science, engineering and technology [49]. Electric machines are used in a broad power range from mWs (micromachines) to 1.7 GWs (4-pole hydrogen/water-cooled turbogenerators). To increase the reliability and simplify maintenance of electromechanical drives, the d.c. commutator machine has gradually been replaced by an energy efficient cage induction motor (IM) and permanent magnet brushless motor (PMBM). PMBMs fall into the two principal classes of synchronous (sinusoidally excited) motors and square wave (trapezoidally excited) motors [68]. There are numerous reasons why electric machines indexmachines!electric continue to naturally grow and evolve, key among them are:

- progress in material engineering : sintered and bonded NdFeB permanent magnets (PMs), high temperature superconductors (HTS), amorphous laminations, cobalt alloy laminations, powder materials, high temperature ferromagnetic alloys, piezoelectric ceramics, magnetostrictive alloys with 'giant' strains, magnetic shape memory (MSM) alloys, wires with heat activated adhesive overcoats, very thin insulation materials, high temperature insulation materials, magnetorheological fluids, etc.;

- new areas of applications e.g., electric vehicles (EVs), hybrid electric vehicles (HEVs), airborne apparatus, space stations, robotics, vacuum, high pressure liquids, harsh environment, nuclear technology, microelectromechanical systems (MEMS), computers, consumer electronics, large drives, bullet trains, marine vessels, submarines, weapon systems, linear metro, magnetic levitation trains;
- impact of power electronics : variable speed drives (VSD), motors with built-in controllers, switched reluctance motor (SRM) drives;
- new control strategies: self-tuning electromechanical drives, 'intelligent' drives, fuzzy control, sensorless control;
- need for energy savings and environmental issues, for example, reduction of pollution and recycling;
- impact of new energy sources, e.g., photovoltaic cells and modules, fuel cells, modern electrochemical batteries, etc.;
- demand on large power, high speed machines, e.g., directed energy weapons (DEWs);
- demand on high torque gearless motor drives, e.g., EVs, light electric traction, machine roomless elevators with direct electromechanical drives [68];
- reliability demands: elimination of brush sliding contacts, fault-tolerant motors, increased time of trouble-free operation;
- new topologies: written pole motors, transverse flux motors (TFMs), hybrid motors, piezoelectric motors, resonant motors, superconducting motors, oscillatory motors, rotary-linear motors, etc.;
- integrated electromechanical drives (motor, encoder, gears, solid state converter, protection, controller and computer interface packaged in one enclosure);
- application of magnetic bearings;
- bearingless motors;
- increased performance–to–cost ratio;
- reduction of noise, vibration, torque ripple, electromagnetic and RF interference;
- impact of mechatronics, i.e., synergistic integration of mechanical engineering with electronics and intelligent computer control in the design and manufacture of products and processes [198];
- impact of computational electromagnetics: the finite element method (FEM), boundary element method (BEM), edge element method (EEM);
- applications of optimization methods, e.g., artificial neural networks, genetic algorithm, etc.;
- new challenges and large scale research programs, e.g.: all electric ship, electric aircraft and 'more electric aircraft', 'more electric engine' (MEE), hybrid electric vehicles (HEVs), utilization of solar power, electric combat vehicles, smart buildings, biomedical and clinical engineering equipment and instrumentation.

1.1 Why electric machines continue to naturally grow

The electromechanical drive family has the biggest share in the electric and electronic market. Motion control market growth is illustrated in the histogram in Fig. 1.1. Currently, d.c. commutator motor drive sales decline whilst the demand for a.c. motor drives increases substantially every year.

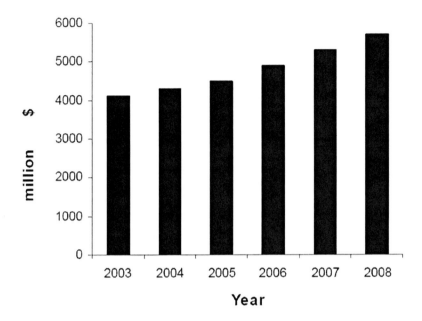

Fig. 1.1. Motion control market growth. Source: *Frost & Sullivan*.

Small PM motors are especially demanded by manufacturers of computer equipment, medical instruments, measurement technology, automobiles, robots, and handling systems.

Advances in electronics, control and PM materials have outpaced similar improvements in associated mechanical transmission systems, making ball screws and gear trains the limiting factors in motion control. On the other hand, linear actuators with ball screws (Fig. 1.2) or roller screws and rotary PM motors have much higher force density than linear PM motors. A substantially higher integration of electromagnetic, electronic and mechanical motor components will increasingly help to bridge this gap in the future. However, there is always the question of the cost, which ultimately is a key factor for specific customer needs.

Table 1.1. Comparison of cage IMs, PMBMs and SRMs rated up to 15 kW and 3600 rpm

Parameter	Cage IM	PMBM	SRM
Power density	Satisfactory	Highest	Lower than PMBM
Maximum shear stress, N/m^2	Up to 20 000	Over 60 000	Up to 35 000
Air gap	Small, fraction of mm	1 mm or more	Small, fraction of mm
Efficiency	Satisfactory; Good for energy efficient motors	Over 90%	About 1% more than that of equivalent IM
Power factor $\cos\phi$	0.8 to 0.9	High, up to 1	Switched d.c. motor
Performance at low speed	Poor	Good	Torque is high, but efficiency low
Torque–voltage characteristics $T = f(V)$	$T \propto V^2$	$T \propto V$	$T \propto V$ at constant peak current
Acoustic noise, dB(A)	Below 60	Below 65	70 to 82
Torque ripple	Less than 5%	Up to 10%	15 to 25%
Overload capacity factor T_{max}/T_{rated}	1.6 to 3.2	About 2	Highest
Power electronics converter	Not necessary for constant speed motor	Necessary	Necessary
Cost	Cost effective motor	More expensive than equivalent IM	Cost effective motor

1.2 Status of electric motors

Comparison of power frequency (50 or 60 Hz) cage IMs, PMBMs and SRMs rated up to 15 kW is shown in Table 1.1. In general, IMs have poor performance at low speed. PMBMs are the highest power density and highest efficiency motors. The SRM technology is still not as mature as other traditional types of electric machines. There are few mass production applications of SRMs so far, e.g., *Maytag®* washing machines.

1.2.1 A.c. motors

Cage IMs have been the most popular electric motors in the 20th Century. Recently, owing to the progress made in the field of power electronics and

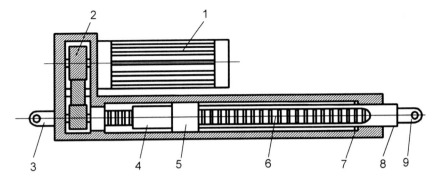

Fig. 1.2. Ball screw linear actuator with rotary PMBM: 1 — PMBM or other electric motor, 2 — gears, 3 — mounting kit, e.g., clevis, 4 — high precision safety ball nut, 5 — internally guided extension tube with anti-rotation mechanism which also acts as a screw support, 6 — ball screw transmission, 7 — extension tube seal, 8 — extension tube, 9 — clevis or sphearical joint end.

control technology, their application to electromechanical drives has significantly increased. The rated output power ranges from 70 W to 500 kW (up to 19 MW for ship propulsion), with 75% of them designed with four–pole stators. The main advantages of cage IMs (Fig. 1.3) are their simple construction, low price, simple maintenance, no commutator or slip rings, low torque ripple and low sound power level. The drawbacks are their small air gap, poor performance at low speeds, torque proportional to the voltage squared, possibility of cracking the rotor bars due to hot spots at plugging and reversal, sensitivity to voltage drop and lower efficiency and power factor than those of synchronous motors. Over 10% of applications use some type of electronic controller either in the form of solid state soft starters or frequency inverters.

Synchronous motors have several advantages in comparison with IMs such as controllable power factor, proportionality between the torque and input voltage, speed dependent only on the input frequency and number of poles, larger air gap and better adaptation to pulsating load torque.

Synchronous motors can operate with unity power factor and even deliver the reactive power to the supply system (power factor correction). Their drawback is much higher price than that of IMs and maintenance of the brush sliding contact in the case of electromagnetic brush exciter. The largest synchronous motors for refrigeration compressors in liquefied natural gas (LNG) plants are rated at 32 to 80 MW and 3600 rpm [102]. The largest synchronous motors for passenger vessel propulsion are rated at 44 MW and 144 rpm (*Queen Elizabeth 2*). A large synchronous motor rated at 55 MW was installed by *ABB* in *Sasol* plant, Secunda, South Africa, in 2002. Synchronous motors with rare-earth PM excitation are the most efficient classical motors and have the highest power density (output power-to-mass or output power-to-volume) or the highest torque density.

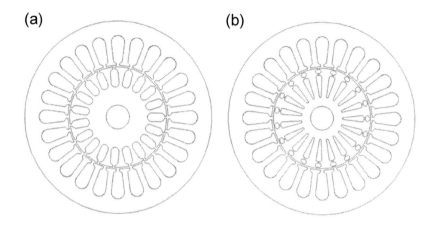

Fig. 1.3. Magnetic circuits of cage induction motors: (a) single cage; (b) double cage. Images have been produced with the aid of SPEED software, University of Glasgow, U.K.

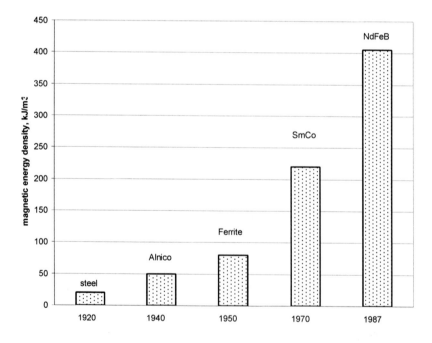

Fig. 1.4. Magnetic energy density of PM materials.

1.2.2 Brushless PM motors

Availability of high energy density rare earth SmCo in the 1970s and NdFeB since 1983 was a breakthrough in PM machine technology and their perfor-

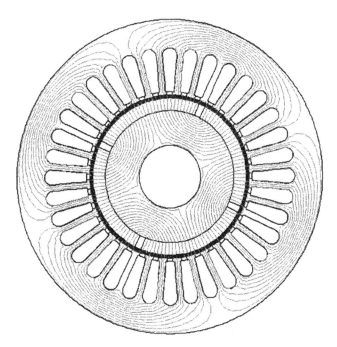

Fig. 1.5. Four-pole PMBM with 36 stator slots. Image has been produced with the aid of SPEED software, University of Glasgow, U.K.

mance. Today, the energy density of NdFeB PMs exceeds 400 kJ/m^3 (Fig. 1.4). Rare–earth PMs improve the output power-to-mass ratio, efficiency, dynamic performance, and reliability [68]. The prices of rare–earth PMs are also dropping, which is making these motors more popular. A PMBM has the magnets mounted on the rotor and the armature winding mounted on the stator (Fig. 1.5). In a d.c. commutator motor the power losses occur mainly in the internal rotor with the armature winding which limits the heat transfer through the air gap to the stator and consequently the armature winding current density (Table 1.2). In PMBMs, all power losses are practically dissipated in the stator where heat can be easily transferred through the ribbed frame or, in larger machines, liquid cooling systems, e.g., water jackets can be used [123].

The PMBM motor shows more advantages than its induction or synchronous reluctance counterparts in motor sizes up to 10 – 15 kW (Table 1.1). The largest commercially available motors are rated at least at 750 kW (1000 hp). There have also been successful attempts to build rare-earth PMBMs rated above 1 MW in Germany and 36.5 MW PMBM by *DRS Technologies*, Parsippany, NJ, U.S.A.

The armature winding of PMBMs is usually distributed in slots. When cogging (detent) torque needs to be eliminated, slotless windings are used. In comparison with slotted windings, the slotless windings provide higher

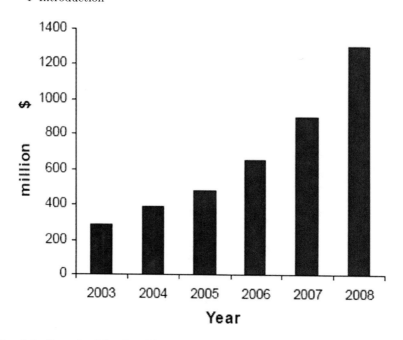

Fig. 1.6. Growth of d.c. brushless motor market. Source: *ARC Advisory Group*.

Table 1.2. Comparison of PM d.c. brushed (commutator) an brushless motors.

	Brushed d.c. motor	Brushless d.c. motor
Commutator (inverter)	Mechanical commutator	Electronic commutator
Maintenance	Commutator and brushes need periodic maintenance	Minimal maintenance
Reliability	Low	High
Moment of Inertia	High	Can be minimized
Power Density	Medium	High
Heat dissipation	Poor (rotor armature winding)	Good (stator armature winding)
Speed control	Simple (armature rheostat or chopper)	Solid state converter required

efficiency at high speeds, lower torque ripple and lower acoustic noise. On the other hand, slotted motors provide higher torque density, higher efficiency in lower speed range, lower armature current and use less PM material.

Participation of PMBMs (Fig. 1.6) in the overall electric motor market is soaring.

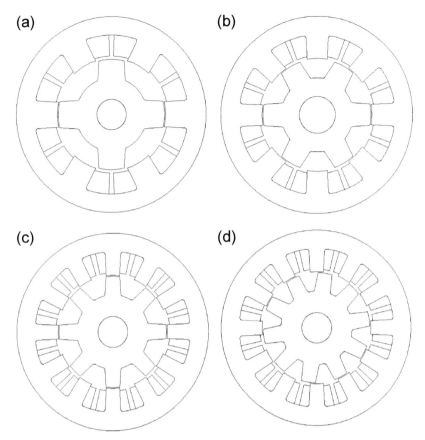

Fig. 1.7. Outlines of of SRMs: (a) three-phase with 6 stator poles and 4 rotor poles; (b) four-phase with 8 stator poles and 6 rotor poles; (c) three or six-phase with 12 stator poles and 8 rotor poles; (d) three or six-phase with 12 stator poles and 10 rotor poles. Images have been produced with the aid of SPEED software, University of Glasgow, U.K.

An auxiliary d.c. field winding helps to increase the speed range over constant power region in hybrid PMBMs or control the output voltage of variable speed generators.

1.2.3 Stepping motors

The *stepping motor* is a singly-excited motor converting electric pulses into angular displacements. It has salient poles both on the stator and rotor and polyphase stator winding. Stepping motors are classified as motors with active rotors (PM rotors), motors with reactive rotors (reluctance type) and hybrid motors (windings, PMs and variable reluctance magnetic circuit). A two-phase hybrid stepping motor performing 200 steps per revolution is nowadays a

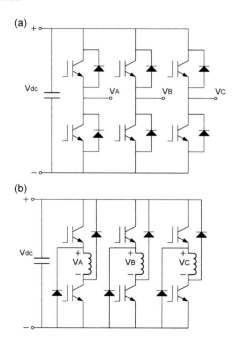

Fig. 1.8. Comparison of solid state converters for three phase motors: (a) IM (inverter), (b) SR motor.

popular motor in factory automation systems. Very high resolution can be achieved in the so called *microstepping mode* in which the basic motor step is subdivided by proportioning the current in the two-phase windings.

1.2.4 Switched reluctance motors

The *switched reluctance motor* (SRM) is a polyphase doubly-salient electric motor with no winding and no excitation system on the rotor. The electromagnetic torque is very sensitive to the *turn-on angle* and *turn-off angle* of the stator phase current. Thus, the SRM controller requires information about the rotor position. The fundamental difference between a stepping reluctance motor and SRM is that the second one requires rotor position sensors. In a three-phase SRM the number of stator poles can be 6 (one phase winding per pole pair) or 12 (one phase winding per two pole pairs) while the number of rotor poles should be 4 or 8, respectively (Fig. 1.7a,c,d). For a four-phase SRM, the number of the stator poles is typically 8 and the number of rotor poles is 6 (Fig. 1.7b).

According to manufacturers, SRMs can provide the highest performance-to-cost ratio, have perspectives of applications in energy efficient drives, high speed drives and fault tolerant drives. Their main drawback is the need for keeping a small air gap and suppressing the torque ripple and acoustic noise. The acoustic noise can be reduced by the use of profiled phase voltage, current or magnetic flux waveforms, mechanical techniques or modified pole geometries. Power electronics converters for SRMs are similar to inverters (Fig. 1.8). The switching frequency is lower and current sharing between solid state devices is better in the case of SRMs. Standard speed SRMs rated below 10 kW are now in mass production, e.g., *Emerson*, St. Louis, MI, U.S.A. Hot market for SRMs comprises washers, dryers, blowers, compressors, servo systems, aircraft and automobile integrated starter-generators, aircraft fuel pump motors. Comparison of cage IMs, PMBMs and SRMs rated up to 15 kW is shown in Table 1.1.

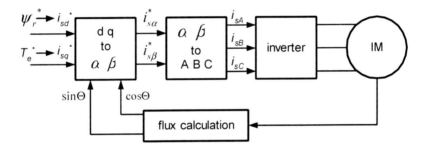

Fig. 1.9. Field oriented control of IM. d–q is the synchronously rotating reference frame, α–β is the stationary reference frame, Ψ_r is the rotor linkage flux, T_e is the electromagnetic torque, i_{sd}, i_{sq} are stator currents in the synchronously rotating reference frame, $i_{s\alpha}$, $i_{s\beta}$ are stator currents in the stationary reference frame, i_{sA}, i_{sB}, i_{sC} are stator phase currents, Θ is rotor flux position angle. Asterisks denote reference currents, flux and torque, respectively.

1.2.5 Servo motors

Servo motor technology has changed in recent years from brushed d.c. and two-phase a.c. servo motor drives to new maintenance-free brushless three-phase vector-controlled a.c. drives for all motor applications where quick response, light weight and large continuous and peak torques are required. The *vector control* is also frequently referred to as *field orientation*. The principle of field oriented control of IM is explained in Fig. 1.9. The equations for a cage IM in the d–q synchronously rotating reference frame are written under assumptions that the air gap is uniform, slot openings are neglected, stator current system is balanced, air gap MMF is sinusoidally distributed and the magnetic saturation is constant (IM operates at fixed point).

Field orientation is a technique that provides decoupling the two components of the stator current: one component producing the air gap magnetic

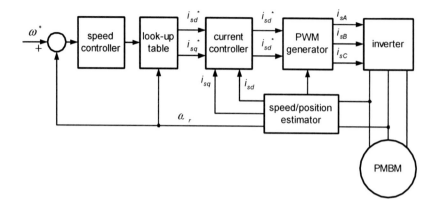

Fig. 1.10. Field oriented control of PM. Description as in Fig. 1.9.

flux and the other producing the torque. Field orientation techniques can be broadly classified into two groups: direct control and indirect control. The *indirect field-oriented control*, suggested by K. Hasse [82] in 1969, requires a high-resolution rotor position sensor, such as an encoder or resolver, to determine the rotor flux position. The direct field-oriented control, as proposed by F. Blaschke in 1972 [27], determines the magnitude and position of the rotor flux vector by direct flux measurement or by a computation based on machine terminal conditions. The IM dynamics lose most of their complexity and a high performance drive can be achieved. The IM drive can also achieve four-quadrant operation with a fast torque response and good performance down to zero speed. On the other hand, the field oriented technique is an excessively complicated control approach requiring sophisticated signal processing and complex transformation. The lengthy computation problems have been overcome with availability of powerful microcomputers.

The block diagram of a field oriented control of PMBM is shown in Fig. 1.10 [214]. The currents are regulated by proportional and integral (PI) regulators in the d–q reference frame [209].

1.3 Progress in electric machines technology

The number of types of electric machines available on the market is in proportion to the technology advancement. In the last three decades many new motors have been developed, e.g., SR motors, piezoelectric motors, transverse flux motors (TFMs), written pole motors, various hybrid motors, PMBMs with nonoverlap stator windings, smart motors, PM linear motors, etc. Synchronous generators and motors with superdonducting (SC) field excitation systems have already entered the prototype stage and very likely can become commercial products in the near future.

1.4 Mechatronics

The term *mechatronics* was first introduced in 1969 by T. Mori of Japanese *Yaskawa* company. Until the early 1980s, mechatronics meant a mechanism that is 'electrified'. In the mid 1980s, mechatronics was seen in Europe as a new branch of engineering bridging the gap between mechanics and electronics. The United States was reluctant to accept the term 'mechatronics' and preferred to use 'system engineering'.

Mechatronics has been defined in many different ways over the years. Definitions include 'incorporating electronics more and more into mechanisms', and 'the integration of mechanical engineering with electronics and intelligent computer control in the design'. In essence, mechatronics means adding an intelligence to the mechanical design or replacing mechanical designs with an intelligent electronic solution. With the technology advancement, designs that were once purely mechanical are now best done with electronics or a combination of both.

Mechatronics gained legitimacy in academic circles in 1996 with the publication of the first referenced journal, IEEE/ASME Transactions on Mechatronics. The authors of the premier issue worked to define mechatronics [19, 109]. After acknowledging that many definitions have circulated, they selected the following: *The synergistic integration of mechanical engineering with electronics and intelligent computer control in the design and manufacturing of industrial products and processes*. Similar definition can be found in the textbook [3]. The term mechatronics is used to denote a rapidly developing, interdisciplinary field of engineering that deals with the design of products, which function relies on the synergistic integration of mechanical, electrical, and electronic components connected by a control architecture. Other definitions can be found, e.g., in [25] and [201, 202, 203].

Today, the term mechatronics encompasses a large array of technologies. Each technology still has the basic element of the merging of mechanics and electronics, but may involve much more, particularly software and information technology.

1.5 Microelectromechanical systems

Microelectromechanical systems (MEMS) are small integrated devices or systems that combine electrical and mechanical components. The range in size is from the sub micrometer (or sub micron) level to the millimeter level. There can be any number of MEMS, from a few to millions, in a particular system. MEMS extend the fabrication techniques developed for the integrated circuit industry to add mechanical elements such as beams, gears, diaphragms, springs and even motors and generators to devices.

Based on semiconductor processing technology, the MEMS promises micrometer range sensors and actuators. While academic researchers in numerous publications have presented successful results in this area, the reality is that the physics of the micro world are not necessarily the same as those of the macro world and the cost of manufacturing MEMS products is higher than that of traditional motor technologies [143]. Other aggravating problems are that very small motors are difficult to interface with other components, have very little usable power, and are tremendously inefficient. They sometimes require very high voltage (100 to 300 V dc) to operate and are very difficult to assemble.

Nevertheless, working prototypes of hybrid MEMS motors are being produced today and some versions will probably soon be available to customers on a limited basis.

1.6 Superconductivity

Superconductivity (SC) is a phenomenon occurring in certain materials at low temperatures, characterized by the complete absence of electrical resistance and the damping of the interior magnetic field (Meissner effect). The *critical!temperature* for SCs is the temperature at which the electrical resistivity of an SC drops to zero. Some critical temperatures of metals are: aluminum (Al) $T_c = 1.2$ K, tin (Sn) $T_c = 3.7$ K, mercury (Hg) $T_c = 4.2$ K, vanadium (V) $T_c = 5.3$ K, lead (Pb) $T_c = 7.2$ K, niobium (Nb) $T_c = 9.2$ K. Compounds can have higher critical temperatures, e.g., $T_c = 92$ K for $YBa_2Cu_3O_7$ and $T_c = 133$ K for $HgBa_2Ca_2Cu_3O_8$. Superconductivity was discovered by Dutch scientist H. Kamerlingh Onnes in 1911 (Nobel Prize in 1913). Onnes was the first person to liquefy helium (4.2 K) in 1908.

The superconducting state is defined by three factors (Fig. 1.11):

- critical temperature T_c;
- critical magnetic field H_c;
- critical current density J_c.

Maintaining the superconducting state requires that both the magnetic field and the current density, as well as the temperature, remain below the critical values, all of which depend on the material.

The phase diagram $T_c H_c J_c$ shown in Fig. 1.11 demonstrates relationship between T_c, H_c, and J_c. When considering all three parameters, the plot represents a critical surface. For most practical applications, superconductors must be able to carry high currents and withstand high magnetic field without everting to their normal state.

Meissner effect (sometimes called Meissner-Ochsenfeld effect) is the expulsion of a magnetic field from a superconductor (Fig. 1.12).

When a thin layer of insulator is sandwiched between two superconductors, until the current becomes critical, electrons pass through the insulator as if it does not exists. This effect is called *Josephson effect*. This phenomenon can be applied to the switching devices that conduct on-off operation at high speed.

1.6 Superconductivity   15

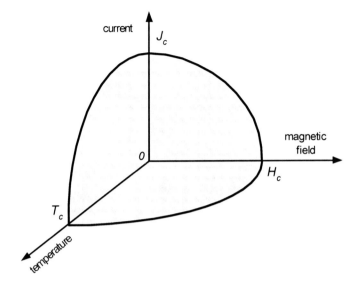

Fig. 1.11. Phase diagram $T_c H_c J_c$.

In *type I superconductors* the superconductivity is 'quenched' when the material is exposed to a sufficiently high magnetic field. This magnetic field, H_c, is called the *critical field*. In constrast, *type II superconductors* have two critical fields. The first is a low-intensity field H_{c1}, which partially suppresses the superconductivity. The second is a much higher critical field, H_{c2}, which totally quenches[1] the superconductivity. The upper critical field of type II superconductors tends to be two orders of magnitude or more above the critical fields of a type I superconductor.

Some consequences of zero resistance are as follows:

- When a current is induced in a ring-shaped SC, the current will continue to circulate in the ring until an external influence causes it to stop. In the 1950s, 'persistent currents' in SC rings immersed in liquid helium were maintained for more than five years without the addition of any further electrical input.
- A SC cannot be shorted out. If the effects of moving a conductor through a magnetic field are ignored, then connecting another conductor in parallel, e.g., a copper plate across a SC will have no effect at all. In fact, by comparison to the SC, copper is a perfect insulator.
- The diamagnetic effect that causes a magnet to levitate above a SC is a complex effect. Part of it is a consequence of zero resistance and of the fact that a SC cannot be shorted out. The act of moving a magnet toward a SC induces circulating persistent currents in domains in the material. These

---

[1] Quenching is a resistive heating of an SC and a sudden temperature rise.

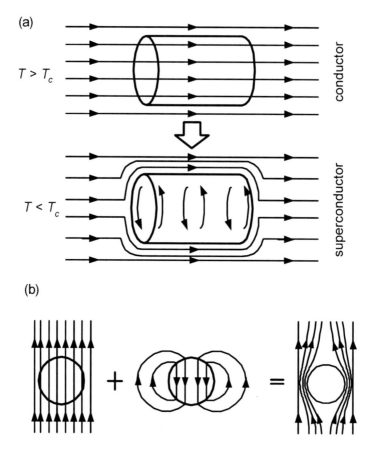

Fig. 1.12. Meissner effect: (a) magnetic field is expelled from SC at $T < T_c$; (b) graphical explanation of Meissner effect.

circulating currents cannot be sustained in a material of finite electrical resistance. For this reason, the *levitating magnet* test is one of the most accurate methods of confirming superconductivity.
- Circulating persistent currents form an array of electromagnets that are always aligned in such as way as to oppose the external magnetic field. In effect, a mirror image of the magnet is formed in the SC with a North pole below a North pole and a South pole below a South pole.

The main factor limiting the field strength of the conventional (Cu or Al wire) electromagnet is the I^2R power losses in the winding when sufficiently high current is applied. In an SC, in which $R \approx 0$, the I^2R power losses practically do not exist.

Lattice (from mathematics point of view) is a partially ordered set in which every pair of elements has a unique *supremum* (least upper bound of elements, called their *join*) and an *infimum* (greatest lower bound, called their *meet*). Lattice is an infinite array of points in space, in which each point has identical surroundings to all others. *Crystal structure* is the periodic arrangement of atoms in the crystal.

The only way to describe superconductors is to use quantum mechanics. The model used is the *BSC theory* (named after Bardeen, Cooper and Schrieffer) was first suggested in 1957 (Nobel Prize in 1973) [20]. It states that:

- lattice vibrations play an important role in superconductivity;
- electron-phonon interactions are responsible.

Photons are the quanta of electromagnetic radiation. *Phonons* are the quanta of acoustic radiation. They are emitted and absorbed by the vibrating atoms at the lattice points in the solid. Phonons possess discrete energy ($E = h\upsilon$) where $h = 6.626\ 068\ 96(33) \times 10^{-34}$ Js is *Planck constant* and propagate through a crystal lattice.

Low temperatures minimize the vibrational energy of individual atoms in the crystal lattice. An electron moving through the material at low temperature encounters less of the impedance due to vibrational distortions of the lattice. The Coulomb attraction between the passing electron and the positive ion distorts the crystal structure. The region of increased positive charge density propagates through the crystal as a quantized sound wave called a phonon. The phonon exchange neutralizes the strong electric repulsion between the two electrons due to Coulomb forces. Because the energy of the paired electrons is lower than that of unpaired electrons, they bind together. This is called *Cooper pairing*. Cooper pairs carry the supercurrent relatively unresisted by thermal vibration of the lattice. Below T_c, pairing energy is sufficiently strong (Cooper pair is more resistant to vibrations), the electrons retain their paired motion and upon encountering a lattice atom do not scatter. Thus, the electric resistivity of the solid is zero. As the temperature rises, the binding energy is reduced and goes to zero when $T = T_c$. Above T_c a Cooper pair is not bound. An electron alone scatters (collision interactions) which leads to ordinary resistivity. Conventional conduction is resisted by thermal vibration within the lattice.

In 1986 J. Georg Bednorz and K. Alex Mueller of IBM Ruschlikon, Switzerland, published results of research [22] showing indications of superconductivity at about 30 K (Nobel Prize in 1987). In 1987 researchers at the University of Alabama at Huntsville (M. K. Wu) and at the University of Houston (C. W. Chu) produced ceramic SCs with a critical temperature ($T_c = 52.5$ K) above the temperature of liquid nitrogen.

There is no widely-accepted temperature that separates *high temperature superconductors* (HTS) from *low temperature superconductors* (LTS). Most LTS superconduct at the boiling point of liquid helium (4.2 K = -269^0C at 1 atm). However, all the SCs known before the 1986 discovery of the superconducting oxocuprates would be classified LTS. The barium-lanthanum-cuprate Ba-La-Cu-O fabricated by Mueller and Bednorz, with a $T_c = 30$ K = -243^0C,

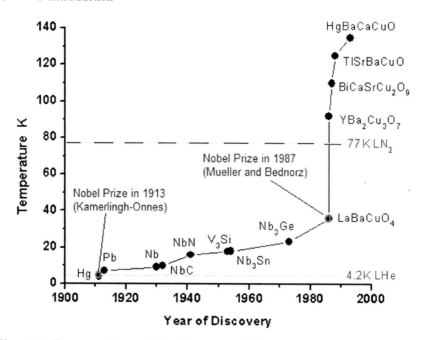

Fig. 1.13. Discovery of materials with successively higher critical temperatures over the last century.

is generally considered to be the first HTS material. Any compound that will superconduct at and above this temperature is called HTS. Most modern HTS superconduct at the boiling point of liquid nitrogen (77 K = -196^0C at 1 atm).

The most important market for LTS superconducting electromagnets are currently *magnetic resonance imaging* (MRI) devices, which enable physicians to obtain detailed images of the interior of the human body without surgery or exposure to ionizing radiation.

All HTS are *cuprates* (copper oxides). Their structure relates to the *perovskite structure* (calcium titanium oxide $CaTiO_3$) with the general formula ABX_3. Perovskite $CaTiO_3$ is a relatively rare mineral occurring in orthorhombic (pseudocubic) crystals[2].

With the discovery of HTS in 1986, the U.S. almost immediately resurrected interest in superconducting applications. The U.S. Department of Energy (DoE) and Defense Advanced Research Projects Agency (DARPA) have taken the lead in research and development of electric power applications. At 60 to 77 K (liquid nitrogen) thermal properties become more friendly and cryogenics can be 40 times more efficient than at 4.2 K (liquid helium).

---

[2] Perovskite (In German 'Perovskit') was discovered in the Ural mountains of Russia by G. Rose in 1839 and named for Russian mineralogist, L. A. Perovski (1792-1856).

In power engineering, superconductivity can be practically applied to synchronous machines, homopolar machines, transformers, energy storages, transmission cables, fault-current limiters, linear synchronous motors and magnetic levitation vehicles. The use of superconductivity in electrical machines reduces the excitation losses, increases the magnetic flux density, eliminates ferromagnetic cores, and reduces the synchronous reactance (in synchronous machines).

Do *room-temperature superconductors* exist? In 1987, a number of researchers believed they had evidence of room-temperature SC. The evidence for this was usually the observation of an instantaneous drop in resistivity as the material is cooled from above room temperature. It is generally believed that the only logical explanation of such a sudden drop in resistivity is that a small filament in the material is superconducting at higher temperature and suddenly 'shunts' part of the non-superconducting material as the filament reaches its critical temperature.

1.7 Solid state converters

Switching capabilities of currently available *thyristors*, *gate turn-off thyristors* (GTOs) and *insulated-gate bipolar transistors* (IGBTs) for high power applications are shown in Fig. 1.14 [212]. IGBTs are now ousting GTO thyristors even at higher range of power and they have already replaced *bipolar junction transistors* (BJTs) [212]. IGBTs will be the main choice for railway traction application in the nearest future [212]. High power thyristors are key solid state devices in the maximum power range.

Both *metal oxide semiconductor field effect transistors* (MOSFETs) and IGBTs developed in the late 1970s and early 1980s are:

- voltage controlled devices as they can be turned on and off by controlling the voltage across their gate-source (emitter) junction;
- have high input impedances;
- have high current carrying capacities.

IGBTs have the following limitations:

- lower switching speeds than MOSFETs as a result of tail current effect (Fig. 1.15a);
- higher switching losses than MOSFET (Fig. 1.15b);
- possibility of uncontrollable latch-up under overstress conditions (high dv/dt or di/dt).

Comparison of bipolar junction transistors (BJTs), IGBTs and MOSFETs is shown in Table 1.3.

IGBTs have been the preferred device under low duty cycle, low frequency < 20 kHz, narrow or small line or load variations, high-voltage applications > 600 V, operation at high junction temperature ($> 100^0$ C), output power

Table 1.3. Comparison of BJTs, IGBTs and MOSFETs. Original data *Power Designers*, Madison, WI, U.S.A.

Criterion	BJT	IGBT	MOSFET
Drive method	Current	Voltage	Voltage
Drive circuit complexity	High	Low	Low
Switching speeds	Slow, μs	Medium	Very fast, ns
Switching frequencies	Few kHz	< 50 kHz	$\leq$ 1 MHz
Forward voltage drop	Low	Low	Medium
Current carrying capability	High	High	Medium
Switching losses	Medium to high	Low to medium	Very low
Breakdown voltage	High	Very high $\approx$ 5000 V	Medium < 1500 V

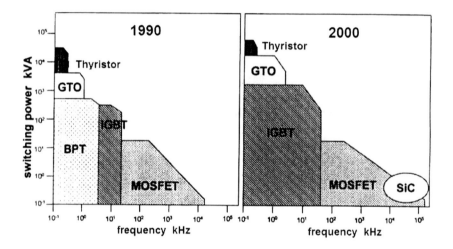

Fig. 1.14. Progress in solid state devices [212].

> 5 kW. Typical IGBT applications include motor control (frequency < 20 kHz, short circuit/in-rush limit protection), constant load low frequency uninterruptible power supply (UPS), welding (high average current, low frequency < 50 kHz, ZVS circuitry), low-power lighting (frequency < 100 kHz). On the other hand, MOSFETs are preferred in high frequency applications > 200 kHz, wide line or load variations, long duty cycles, low-voltage applications < 300 V, output power < 10 kW. Typical MOSFET applications are switch mode power supplies (SMPS) with hard switching above 200 kHz, SMPS with zero voltage switching (ZVS) rated below 1000 W, battery charging.

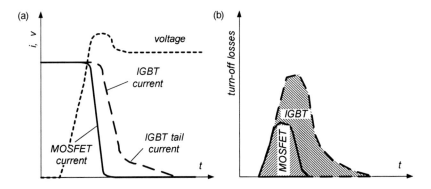

Fig. 1.15. Turn-off behaviour of MOSFET and IGBT: (a) currents and voltage versus time; (b) turn-off losses versus time.

The *integrated gate commutated thyristor* (IGCT) is the optimum combination of the low-loss thyristor technology and the snubberless, cost-effective GTO for medium and high voltage power electronics applications [212, 182].

Silicon will remain as the basic material for solid state devices in the next 10 to 15 years. *Silicon carbide* (SiC) and diamond have been suggested as suitable materials for the construction of new generation of semiconductor devices. SiC can be used in harsh conditions up to 600^0C.

The volume of water cooled solid state converters decreases by about 80% at 150 kW as compared with standard air cooled converters of the same size (*Rockwell Automation*). In addition, cost decreases by 20% versus optimized air cooled converters. Fig. 1.16 shows the construction of a water cooled converter developed by *Rockwell Automation (Reliance Electric)*, Cleveland, IL, U.S.A. The tubing system is embedded into a cast aluminum heat sink. Solid stated devices and capacitors are fixed directly to the heat sink. Control circuitry boards are attached to the front metal panel.

An example of large-power solid state converter is shown in Fig. 1.17. This high performance medium voltage converter is based on powerful press-pack IGBTs, often eliminates requirements for filters and never requires power factor correction [149]. The MV7000 family covers the medium and high power range up to 32 MW at motor voltages of 3.3 to 6.6 kV.

The a.c. pulse-width modulation (PWM) variable speed drive (VSD) is now a standard item, with nothing likely to be a serious competitor in the volume market in the foreseeable future.

1.8 Energy conservation

The contemporary world faces enormous increase in energy consumption and drastic pollution of our planet. Politicians and decision makers should take

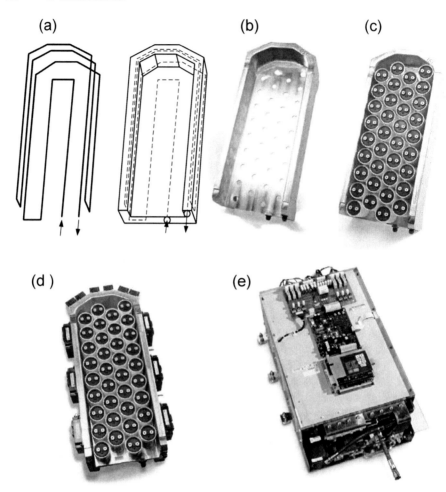

Fig. 1.16. Construction of a liquid cooled solid state converter developed by *Rockwell Automation*, Cleveland, OH, U.S.A.: (a) formed swaged copper tubing, (b) cast aluminum heat sink, (c) heat sink with capacitors, (d) heat sink with capacitors, solid state devices and diodes, (e) assembled converter with regulator and control boards. Courtesy of *Rockwell Automation*, Cleveland, IL, U.S.A.

into consideration the opinions of scientists and engineers how to effectively minimize these problems.

The total electricity consumed annually plus imports and minus exports, expressed in kWh is given in Table 1.4. The *world electricity consumption* was estimated at about 12 000 TWh per annum in 1996, 16 280 in 2006 and it is predicted to increase to about 19 000 TWh in 2010 and about 22 000 TWh in 2020 [68]. The industrial sector, in developed countries, uses more than 30% of the electrical energy. More than 65% of the electrical energy is consumed by electric motor drives. It has been estimated, that in developed industrialized

Fig. 1.17. Large 32 MW MV7000 solid state converter based on IGBTs for induction and synchronous motors. From left to right: control unit, cooling unit, front end, inverter, d.c. cubicle. Photo courtesy of *Converteam*, Massy, France.

countries, up to 10% of electrical energy can be saved by using more efficient control strategies for electromechanical drives. This means that electrical machines have an enormous influence on the reduction of energy consumption. Electrical energy consumption can be saved in one of the following ways: (a) good housekeeping, (b) use of variable-speed drives and (c) construction of electric motors with better efficiency.

Most energy is consumed by three-phase IMs rated below 10 kW. An *energy-efficient IM* produces the same shaft output power, but uses less electrical input power than a standard efficiency motor. Energy efficient IMs have:

- more copper in the winding;
- better quality and thinner steel laminations in the stator;
- smaller air gap between the stator and rotor;
- reduced ventilation (fan) losses;
- closer machining tolerances.

1.9 Power quality

With growing applications of power electronics, *power quality* related issues become more vital. Converter fed motors deteriorate the power quality of the utility system and vice versa – the poor quality of power supply results in degrading the electromechanical drive performance. To improve the power quality, development and application of electromagnetic interference (EMI) filters, converters with EMI suppression components, novel static VAR compensators, PWM active power line conditioners (APLC) and flexible a.c. transmission systems (FACTSs) is more and more important. The ideal scenario is a *green power electronics*, i.e., with no harmonics pollution and unity power factor.

Table 1.4. Electricity Consumption by country in 2006. Source *CIA World Factbook*, 2007.

Rank	Country	Amount, kWh
1	United States	3 717 000 000 000
2	China	2 494 000 000 000
3	Japan	946 300 000 000
4	Russia	940 000 000 000
5	India	587 900 000 000
6	Germany	524 600 000 000
7	Canada	522 400 000 000
8	France	482 400 000 000
9	Brazil	415 900 000 000
10	United Kingdom	345 200 000 000
11	South Korea	321 000 000 000
12	Italy	303 800 000 000
13	Spain	241 800 000 000
14	Mexico	224 600 000 000
15	Australia	209 500 000 000
16	South Africa	207 000 000 000
17	Ukraine	181 900 000 000
18	Taiwan	175 300 000 000
19	Iran	145 100 000 000
20	Saudi Arabia	144 400 000 000
21	Turkey	140 300 000 000
22	Sweden	137 800 000 000
23	Poland	124 100 000 000
24	Thailand	116 200 000 000
25	Norway	112 800 000 000
Total (209 countries)		16 282 774 340 000
Weighted average		77 908 011 196.2

1.10 Recyclable electric machines

After failure or longtime use, when repairing is not economically justified or electric machine is not repairable, it can be handled, generally, within the following categories [111]:

(a) discarded into the environment;
(b) placed in a permitted landfill;
(c) put to a high-value use, breaking it down into its components, melting steel, copper and aluminun;
(d) rebuilt (totally or partially), some componets discarded or reused;
(e) reused.

Since categories (a) and (b) have negative effect on the environment, they are not recycling. Categories (c), (d) and (e) can be classified as recycling, because they create value at the end of life of electric machine.

Design and construction guidelines for *recyclable electric machines* include, but are not limited to:

- the number of parts should be reduced;
- all parts should be made simple;
- in mechanical design both assembly and disassembly should be considered;
- the number of materials should be limited;
- toxic materials, e.g., beryllium copper, lead based alloys, etc., should be avoided;
- usage of recyclable materials should be maximized;
- ferromagnetic, current conductive and insulating materials should not age and, when possible, their performance should improve with time;
- as many dimensions and shapes as possible should be standarized;
- opportunities for using old or rebuilt parts in new machines should be considered.

An ideal machine for recycling is a machine with ferrite PMs, sintered powder magnetic circuit and slotless winding. After crushing powder materials, conductors can be separated and reused. Design and manufacture of recyclable electric machine is economically justified if costs of the final product does not exceed significantly costs of a similar non-recyclable machine.

2
Material engineering

In this chapter modern active materials for construction of electrical machines and electromagnetic devices have been discussed, i.e.,

- laminated silicon steels;
- high saturation alloys;
- amorphous ferromagnetic alloys;
- soft magnetic powder composites;
- PMs;
- high temperature wires;
- insulating materials;
- high temperature superconductors (HTS);
- nanostructured materials.

2.1 Laminated silicon steels

Addition of 0.5% to 3.25% of silicon (Si) increases the resistivity (reduces eddy current losses) and improves magnetization curves B–H of low *carbon steels*. Silicon content , however, increases hardness of laminations and, as a consequence, shortens the life of stamping tooling.

Nonoriented electrical steels are Fe-Si alloys with random orientation of crystal cubes and practically the same properties in any direction in the plane of the sheet or ribbon. Nonoriented electrical steels are available as both *fully processed* and *semi-processed* products. Fully processed steels are annealed to optimum properties by manufacturer and ready for use without any additional processing. Semi-processed steels always require annealing after stamping to remove excess carbon and relieve stress. Better grades of silicon steel are always supplied fully processed, while semi processed silicon steel is available only in grades M43 and worse. In some cases, users prefer to develop the final magnetic quality and achieve relief of fabricating stresses in laminations or assembled cores for small machines.

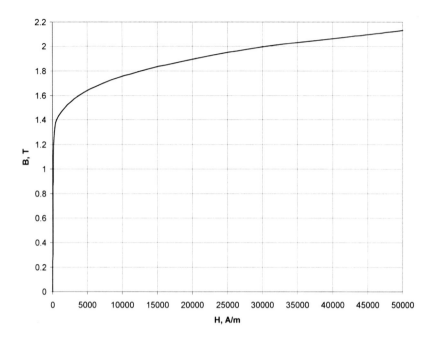

Fig. 2.1. Magnetization curve of fully processed *Armco* DI-MAX nonoriented electrical steel M-19.

Silicon steels are generally specified and selected on the basis of allowable *specific core losses* (W/kg or W/lb). The most universally accepted grading of electrical steels by core losses is the American Iron and Steel Industry (AISI) system (Table 2.1), the so called 'M-grading'. The M number, e.g., M19, M27, M36, etc., indicates maximum specific core losses in W/lb at 1.5 T and 50 or 60 Hz, e.g., M19 grade specifies that losses shall be below 1.9 W/lb at 1.5 T and 60 Hz. The electrical steel M19 offers nearly the lowest core loss in this class of material and is probably the most common grade for motion control products (Fig. 2.1).

The increase of power loss with time or *ageing* is caused by an excessive carbon content in the steel. Modern non-oriented fully processed electrical steels are free from magnetic ageing [195].

The magnetization curve of nonoriented electrical steels M-27, M-36 and M-43 is shown in Fig. 2.2. Core loss curves of nonoriented electrical steels M-27, M-36 and M-43 measured at 60 Hz are given in Table 2.2. Core losses at 50 Hz are approximately 0.79 times the core loss at 60 Hz.

For modern high efficiency, high performance applications, there is a need for operating a.c. devices at higher frequencies, i.e., 400 Hz to 10 kHz. Because of the thickness of the standard silicon ferromagnetic steels 0.25 mm (0.010") or more, core loss due to eddy currents is excessive. Nonoriented electrical

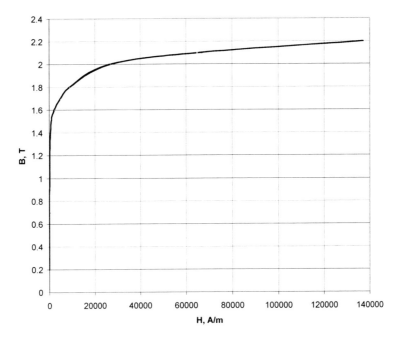

Fig. 2.2. Magnetization curve of fully processed *Armco* DI-MAX nonoriented electrical steels M-27, M-36 and M-43. Magentization curves for all these three grades are practically the same.

steels with thin gauges (down to 0.025 mm thick) for ferromagnetic cores of high frequency rotating machinery and other power devices are manufactured, e.g., by *Arnold Magnetic Technologies Corporation*, Rochester, NY, U.S.A. *Arnold* has two standard nonoriented lamination products: Arnon^{TM} 5 (Figs 2.3 and 2.4) and Arnon^{TM} 7. At frequencies above 400 Hz, they typically have less than half the core loss of standard gauge non-oriented silicon steel laminations.

Fully processed non-oriented electrical steel, suited for use at medium to high frequencies (typically 200 to 3000 Hz) are manufactured by *Cogent Power Ltd.*, Newport, UK [195]. Thin non-oriented grades are available at two nominal thicknesses 0.12 mm (0.005"), 0.18 mm (0.007") and 0.20 mm (0.008"). Typical chemical composition is 3.0% Si, 0.4% Al and 96.6% Fe. The one-side thickness of inorganic phosphate based coating with inorganic fillers and some organic resin (Suralac^{TM} 7000) is 0.001 mm. Specific core losses and magnetization curves of *Cogent* nonoriented thin electrical steels are shown in Table 2.3. The thermal conductivity is 28 W/(m K) in the plane of lamination and 0.37 W/(m K) normal to the plane of lamination. The specific mass density is 7650 kg/m^3, Young modulus 185 GPa and electric conductivity 1.92×10^6 S/m.

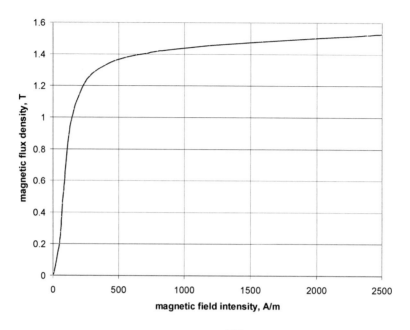

Fig. 2.3. Magnetization curve of Arnon^{TM} 5 nonoriented electrical steel.

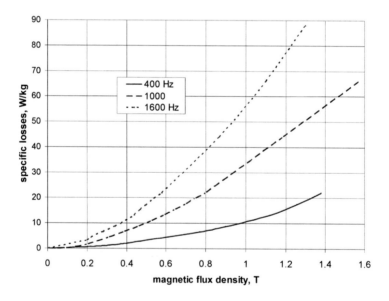

Fig. 2.4. Specific core loss curves of Arnon^{TM} 5 nonoriented electrical steel.

Table 2.1. Silicon steel designations specified by European, American, Japanese and Russian standards

Europe IEC 404-8-4 (1986)	U.S.A. AISI	Japan JIS 2552 (1986)	Russia GOST 21427 0-75
250-35-A5	M 15	35A250	2413
270-35-A5	M 19	35A270	2412
300-35-A5	M 22	35A300	2411
330-35-A5	M 36	—	—
270-50-A5	—	50A270	—
290-50-A5	M 15	50A290	2413
310-50-A5	M 19	50A310	2412
330-50-A5	M 27	—	—
350-50-A5	M 36	50A350	2411
400-50-A5	M 43	50A400	2312
470-50-A5	—	50A470	2311
530-50-A5	M 45	—	2212
600-50-A5	—	50A600	2112
700-50-A5	M 47	50A700	—
800-50-A5	—	50A800	2111
350-65-A5	M19	—	—
400-65-A5	M27	—	—
470-65-A5	M43	—	—
530-65-A5	—	—	2312
600-65-A5	M45	—	2212
700-65-A5	—	—	2211
800-65-A5	—	65A800	2112
1000-65-A5	—	65A1000	—

2.2 High saturation ferromagnetic alloys

Iron–cobalt (Fe–Co) alloys with Co contents ranging from 15 to 50% have the highest known saturation magnetic flux density, about 2.4 T at room temperature. They are the natural choice for applications such as aerospace (motors, generators, transformers, magnetic bearings) where mass and space saving are of prime importance. Additionally, the iron-cobalt alloys have the highest Curie temperatures of any alloy family and have found use in elevated temperature applications. The nominal composition, e.g., for Hiperco 50 from *Carpenter*, PA, U.S.A. is 49% Fe, 48.75% Co, 1.9% V, 0.05% Mn, 0.05% Nb and 0.05% Si. Hiperco 50 has the same nominal composition as Vanadium Permendur and Permendur V.

The specific mass density of Hiperco 50 is 8120 kg/m^3, modulus of elasticity 207 GPa, electric conductivity 2.5×10^6 S/m, thermal conductivity 29.8 W/(m K), Curie temperature 940^{0}C, specific core loss about 76 W/kg at 2 T, 400 Hz and thickness from 0.15 to 0.36 mm. The magnetization curve of

Table 2.2. Specific core losses of *Armco* DI-MAX nonoriented electrical steels M-27, M-36 and M-43 at 60 Hz

Magnetic flux density T	Specific core losses W/kg							
	0.36 mm		0.47 mm			0.64 mm		
	M-27	M-36	M-27	M-36	M-43	M-27	M-36	M-43
0.20	0.09	0.10	0.10	0.11	0.11	0.12	0.12	0.13
0.50	0.47	0.52	0.53	0.56	0.59	0.62	0.64	0.66
0.70	0.81	0.89	0.92	0.97	1.03	1.11	1.14	1.17
1.00	1.46	1.61	1.67	1.75	1.87	2.06	2.12	2.19
1.30	2.39	2.58	2.67	2.80	2.99	3.34	3.46	3.56
1.50	3.37	3.57	3.68	3.86	4.09	4.56	4.70	4.83
1.60	4.00	4.19	4.30	4.52	4.72	5.34	5.48	5.60
1.70	4.55	4.74	4.85	5.08	5.33	5.99	6.15	6.28
1.80	4.95	5.14	5.23	5.48	5.79	6.52	6.68	6.84

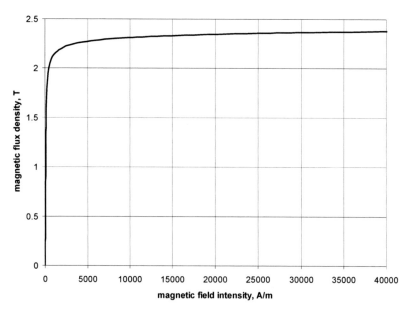

Fig. 2.5. Magnetization curve of Hiperco50.

Table 2.3. Specific core losses and d.c. magnetization curve of nonoriented thin electrical steels manufactured by *Cogent Power Ltd.*, Newport, UK [195].

Magn. flux dens. T	Specific core losses W/kg									Magn. field strength kA/m
	NO 12 0.12 mm			NO 18 0.18 mm			NO 20 0.2 mm			
	50 Hz	400 Hz	2.5 kHz	50 Hz	400 Hz	2.5 kHz	50 Hz	400 Hz	2.5 kHz	
0.10	0.02	0.16	1.65	0.02	0.18	2.18	0.02	0.17	2.79	0.025
0.20	0.08	0.71	6.83	0.08	0.73	10.6	0.07	0.72	10.6	0.032
0.30	0.16	1.55	15.2	0.16	1.50	19.1	0.14	1.49	24.4	0.039
0.40	0.26	2.57	2.54	0.26	2.54	31.7	0.23	2.50	40.4	0.044
0.50	0.37	3.75	37.7	0.36	3.86	45.9	0.32	3.80	58.4	0.051
0.60	0.48	5.05	52.0	0.47	5.22	61.5	042	5.17	78.4	0.057
0.70	0.62	6.49	66.1	0.61	6.77	81.1	0.54	6.70	103.0	0.064
0.80	0.76	8.09	83.1	0.75	8.47	104.0	0.66	8.36	133.0	0.073
0.90	0.32	9.84	103.0	0.90	10.4	161.0	0.80	10.3	205.0	0.084
1.00	1.09	11.8	156	1.07	12.3	198.0	0.95	12.2	253.0	0.099
1.10	1.31	14.1		1.28	14.9		1.14	14.8		0.124
1.20	1.56	16.7		1.52	18.1		1.36	17.9		0.160
1.30	1.89	19.9		1.84	21.6		1.65	21.4		0.248
1.40	2.29	24.0		2.23	25.6		2.00	25.3		0.470
1.50	2.74	28.5		2.67	30.0		2.40	29.7		1.290
1.60	3.14			3.06			2.75			3.550
1.70	3.49			3.40			3.06			7.070
1.80	3.78			3.69			3.32			13.00

Hiperco 50 is shown in Fig. 2.5. Specific losses (W/kg) in 0.356 mm strip at 400 Hz of iron-cobalt alloys from *Carpenter* are given in Table 2.4. Physical properties of Hiperco 50 and electrical steels are compared in Table 2.5. Typical magnetic properties of iron-cobalt alloys from *Carpenter* are given in Table 2.6.

Similar to Hyperco 50 is Vacoflux 50 (50% Co) cobalt-iron alloy from *Vacuumschmelze*, Hanau, Germany, typically used for manufacturing very high flux density pole-shoes, electromagnets with large lifting force, magnetic lenses, needle printers, relays, motors and actuators with high torques and forces [177]. Figs 2.6 and 2.7 show B–H magnetization curves and specific loss curves of Vacoflux 50 and Vacoflux 17. Vacoflux 17 (17% Co) is used for devices and actuators for automotive industry, turned parts and extruded parts.

34   2 Material engineering

Table 2.4. Specific losses (W/kg) in 0.356 mm iron-cobalt alloy strips at 400 Hz.

Alloy	Magnetic flux density		
	1.0 T	1.5 T	2.0 T
Hiperco 15	30	65	110
Permendur 24	42	105	160
Hiperco 27	53	110	180
Rotelloy 5	40	130	200
Hiperco 50	25	44	76
Hiperco 50A	14	31	60
Hiperco 50HS	43	91	158
Rotelloy 3	22	55	78
Permendur 49	22	55	78
Rotelloy 8	49	122	204
HS 50	–	–	375

Table 2.5. Comparison of physical properties of Hiperco 50 with electrical steels.

Property	Hiperco 50	Q-core XL	M15
Saturation flux density, T	2.38	2.10	2.0
Electric conductivity, S/m	2.38	2.94	2.00
Curie temperature, ^{0}C	940	770	760
Maximum relative permeability	10 000	> 14 000	8300
Core losses at 400 Hz and 1.5 T, W/kg	44	57	44
Core losses at 400 Hz and 2.0 T, W/kg	76	–	–
Yield strength, MPa	430	310	360

Table 2.6. Typical magnetic properties of Hiperco alloys.

Alloy	μ_{rmax}	H_{cr} A/m	Normal magnetic flux density, T				
			800 A/m	1600 A/m	4000 A/m	8000 A/m	16 000 A/m
Hiperco 15	1700	200	1.17	1.45	1.73	1.91	2.09
Permendur 24	2000	140	1.50	1.70	1.95	2.00	2.10
Hiperco 27 HS	300	960	0.6	1.20	1.80	2.00	2.15
Hiperco 27	2100	150	1.49	1.72	1.94	2.08	2.26
Rotelloy 5	3000	280	1.89	2.06	2.22	2.25	2.25
Hiperco 50	7000	120	2.14	2.22	2.26	2.31	2.34
Hiperco 50A	25 000	30	2.20	2.25	2.30	2.33	2.35
Hiperco 50HS	2650	410	1.90	2.08	2.21	2.27	2.31
Rotelloy 3	13 000	65	2.15	2.22	2.29	2.34	2.34
Permendur 49	7000	80	2.15	2.22	2.29	2.34	2.34
Rotelloy 8	3300	440	1.70	2.05	2.22	2.27	2.38
HS 50	3800	290	2.17	2.29	2.36	2.38	2.44

2.2 High saturation ferromagnetic alloys 35

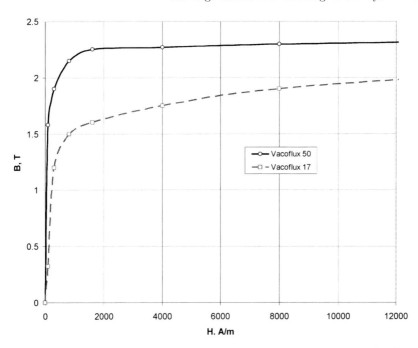

Fig. 2.6. Magnetization curves B–H of Vacoflux 50 and Vacoflux 17 [177].

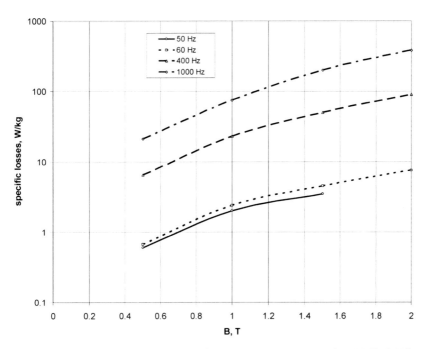

Fig. 2.7. Specific loss curves of Vacoflux 50 at 50, 60, 400 and 1000 Hz [177].

2.3 Amorphous ferromagnetic materials

Core losses can be substantially reduced by replacing standard electrical laminated steels with *amorphous ferromagnetic alloys*. *Metglass* amorphous alloys (*Honeywell (Allied-Signal)*) have specific core losses at 1 T and 50 Hz from 0.125 to 0.28 W/kg.

Amorphous alloy ribbons based on alloys of iron, nickel and cobalt are produced by rapid solidification of molten metals at cooling rates of about 10^6 °C/s. The alloys solidify before the atoms have a chance to segregate or crystallize. The result is a metal alloy with a glass-like structure, i.e., a non-crystalline frozen liquid.

The efficiency of a standard small 550 W induction motor is 74% [8]. It means that power losses dissipated in this motor are 137 W. Replacing the standard core with amorphous alloy core, the losses are reduced to 88 W, so that the efficiency increases to 84% [8]. Application of amorphous ribbons to the mass production of electric motors is limited by their hardness, i.e., 800 to 1100 in Vicker's scale which requires alternative cutting methods as a liquid jet. Standard cutting methods like a guillotine or blank die are not suitable. The mechanically stressed amorphous material cracks. Laser and electrical discharge machining[1] (EDM) cutting methods melt the amorphous material and cause undesirable crystallization. In addition, these methods make electrical contacts between laminations which contribute to the increased eddy-current and additional losses. In the early 1980s chemical methods were used by *General Electric* to cut amorphous materials but these methods were very slow and expensive [122]. The problem of cutting hard amorphous ribbons can be overcome by using a liquid jet [179, 180]. This method makes it possible to cut amorphous materials in ambient temperature without cracking, melting, crystallization and electric contacts between isolated ribbons.

2.4 Soft magnetic powder composites

New *soft magnetic powder materials* which are competitive to traditional steel laminations have recently been developed [1, 91]. Powder materials are recommended for 3D magnetic circuits such as claw-pole, transverse flux (TFMs), disc type and recyclable machines. Specific core losses at 1 T and 100 Hz are 9 W/kg for Accucore (U.S.A.) and 12.5 W for Somaloy 500 (Sweden). At 10 kA/m the magnetic flux density is 1.72 T for Accucore and 1.54 T for Somaloy 500.

Powder metallurgy is used in the production of ferromagnetic cores of small electrical machines or ferromagnetic cores with complicated shapes. The components of *soft magnetic powder composites* are iron powder, dielectric (epoxy

---

[1] EDM works by eroding material in the path of electrical discharges that form an arc between an electrode tool and the work piece.

Table 2.7. Physical properties of iron based METGLAS amorphous alloy ribbons, *Honeywell*, Morristown, NJ, U.S.A.

Quantity	2605CO	2605SA1
Saturation magnetic flux density, T	1.8	1.59 annealed / 1.57 cast
Specific core losses at 50 Hz and 1 T, W/kg	less than 0.28	about 0.125
Specific density, kg/m^3	7560	7200 annealed / 7190 cast
Electric conductivity, S/m	0.813×10^6 S/m	0.769×10^6 S/m
Hardness in Vicker's scale	810	900
Elastic modulus, GN/m^2	100...110	100...110
Stacking factor	less than 0.75	less than 0.79
Crystallization temperature, °C	430	507
Curie temperature, °C	415	392
Maximum service temperature, °C	125	150

Table 2.8. Specific core losses of iron based METGLAS amorphous alloy ribbons, *Honeywell*, Morristown, NJ, U.S.A.

| Magnetic flux density, B T | Specific core losses, Δp, W/kg | | | |
| | 2605CO | | 2605SA1 | |
	50 Hz	60 Hz	50 Hz	60 Hz
0.05	0.0024	0.003	0.0009	0.0012
0.10	0.0071	0.009	0.0027	0.0035
0.20	0.024	0.030	0.0063	0.008
0.40	0.063	0.080	0.016	0.02
0.60	0.125	0.16	0.032	0.04
0.80	0.196	0.25	0.063	0.08
1.00	0.274	0.35	0.125	0.16

resin) and filler (glass or carbon fibers) for mechanical strengthening. Powder composites for ferromagnetic cores of electrical machines and apparatus can be divided into [210]:

- dielectromagnetics and magnetodielectrics,
- magnetic sinters.

38   2 Material engineering

Table 2.9. Magnetization and specific core loss characteristics of non-sintered *Accucore* powder material, *TSC Ferrite International*, Wadsworth, IL, U.S.A.

Magnetization curve		Specific core loss curves		
Magnetic flux density, B T	Magnetic field intensity, H A/m	60 Hz W/kg	100 Hz W/kg	400 Hz W/kg
0.10	152	0.132	0.242	1.058
0.20	233	0.419	0.683	3.263
0.30	312	0.772	1.323	6.217
0.40	400	1.212	2.072	9.811
0.50	498	1.742	2.976	14.088
0.60	613	2.315	3.968	18.850
0.70	749	2.954	5.071	24.295
0.80	909	3.660	6.305	30.490
0.90	1107	4.431	7.650	37.346
1.00	1357	5.247	9.039	44.489
1.10	1677	6.129	10.582	52.911
1.20	2101	7.033	12.214	61.377
1.30	2687	7.981	13.845	70.151
1.40	3525	8.929	15.565	79.168
1.50	4763	9.965	17.394	90.302
1.60	6563	10.869	19.048	99.671
1.70	9035	11.707	20.635	109.880
1.75	10,746	12.125	21.407	

Table 2.10. Specific core losses of Somaloy^{TM} 500 +0.5% Kenolube, 800 MPa, treated at 500°C for 30 min in the air, *Höganäs*, Höganäs, Sweden

Magnetic flux density, T	Specific losses W/kg					
	50 Hz	100 Hz	300 Hz	500 Hz	700 Hz	1000 Hz
0.4	1.5	3	12	18	27	45
0.5	1.9	3.6	17	27	40	60
0.6	2.7	6	21	34	55	90
0.8	4.6	10	32	52	92	120
1.0	6.8	16	48	80	140	180
2.0	30	50	170	270	400	570

2.4 Soft magnetic powder composites

Table 2.11. Magnetization curves of Somaloy^{TM} 500 +0.5% Kenolube, treated at 500°C for 30 min in the air, *Höganäs*, Höganäs, Sweden

Magnetic field intensity H A/m	B at density 6690 kg/m^3 T	B at density 7100 kg/m^3 T	B at density 7180 kg/m^3 T
1 500	0.7	0.83	0.87
3 200	0.8	1.13	1.19
4 000	0.91	1.22	1.28
6 000	1.01	1.32	1.38
10 000	1.12	1.42	1.51
15 000	1.24	1.52	1.61
20 000	1.32	1.59	1.69
40 000	1.52	1.78	1.87
60 000	1.65	1.89	1.97
80 000	1.75	1.97	2.05
100 000	1.82	2.02	2.10

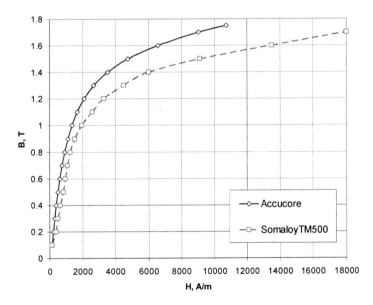

Fig. 2.8. Comparison of magnetization curves of *Accucore* and *Somaloy*^{TM} 500.

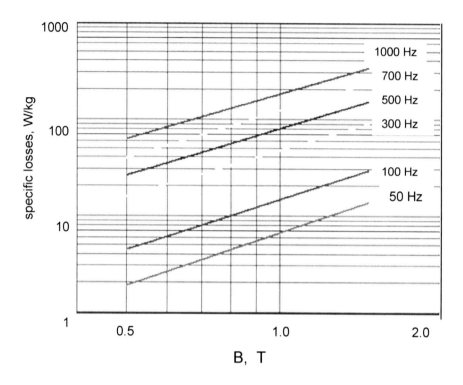

Fig. 2.9. Specific losses as functions of the magnetic flux density at frequencies from 50 to 1000 Hz and temperature 500°C for $Somaloy^{TM}500$ with specific density 7300 kg/m^3 and 0.5% Kenolube.

Dielectromagnetics and magnetodielectrics are names referring to materials consisting of the same basic components: ferromagnetic (mostly iron powder) and dielectric (mostly epoxy resin) material [210]. The main tasks of the dielectric material is insulation and binding of ferromagnetic particles. In practice, composites containing up to 2% (of their mass) of dielectric materials are considered as *dielectromagnetics*. Those with a higher content of dielectric material are considered as *magnetodielectrics* [210].

TSC International, Wadsworth, IL, U.S.A., has developed a new soft powder material, *Accucore*, which is competitive to traditional steel laminations [1]. The magnetization curve and specific core loss curves of the non-sintered *Accucore* are given in Table 2.9. When sintered, *Accucore* has higher saturation magnetic flux density than the non-sintered material. The specific density is 7550 to 7700 kg/m^3.

Höganäs, Höganäs, Sweden, manufactures soft magnetic composite (SMC) powders that are surface-coated metal powders with excellent compressibility [91, 141]. $Somaloy^{TM}500$ (Tables 2.10 and 2.11) has been developed for 3D

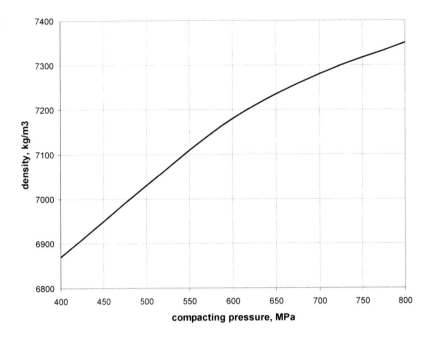

Fig. 2.10. Effect of compacting pressure on the specific mass density of *Höganäs* soft magnetic composite powders.

magnetic circuits of electrical machines, transformers, ignition systems and sensors. Comparison of magnetization characteristics of *Accucore* and *Somaloy*^{TM}500 is shown in Fig. 2.8. Specific losses as function of the magnetic flux density at frequencies from 50 to 1000 Hz and temperature 500^0C for *Somaloy*^{TM}500 with specific density 7300 kg/m^3 and 0.5% Kenolube are shown in Fig. 2.9 [91].

Using SMC powders, e.g., the stator core of an axial flux PM (AFPM) machine can be made as a slotted core, slotless cylindrical core and salient pole core with one coil per pole (non-overlap winding).

The slotted and slotless cores can be made in a *powder metallurgy* process using a ferromagnetic powder with a small amount of lubricants or binders. The powder metallurgy process generally consists of four basic steps, namely: (1) powder manufacture, (2) mixing or blending, (3) compacting and (4) sintering. Most compacting is done with mechanical, hydraulic or pneumatic presses and rigid tools. Compacting pressures generally range between 70 to 800 MPa with 150 to 500 MPa being the most common. The outer diameter of the core is limited by the press capability. Frequently, the stator core must be divided into smaller segments. Most powder metallurgy products must have cross sections of less than 2000 mm^2. If the press capacity is sufficient, sec-

Fig. 2.11. SMC powder salient pole stator for disc type (axial flux) PM motors. Courtesy of *Höganäs*, Höganäs, Sweden.

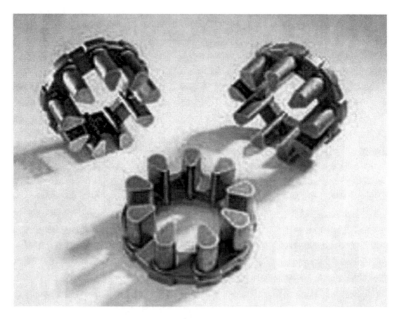

Fig. 2.12. Powder salient pole stators for small single-sided disc type axial flux PM motors. Courtesy of *Mii Technologies, LLC*, Lebanon, NH, U.S.A.

tions up to 6500 mm² can be pressed. Fig. 2.10 shows the effect of compacting pressure on the density of *Höganäs* SMC powders.

For *Somaloy*™500 the heat treatment temperature (sintering) is typically 500°C for 30 min. After heat treatment the compacted powder has much less mechanical strength than solid steel.

The thermal expansion of conductors within the stator slots creates thermal expansion stresses in the stator teeth. The magnitude of these stresses depends upon the difference in the temperature of the winding and core, difference in coefficients of thermal expansion of both materials and slot fill factor. This problem is more important in powder cores than in laminated cores since the tensile stress of powder cores is at least 25 times lower and their modulus of elasticity is less than 100 GPa (versus 200 GPa for steel laminations).

SMC powder salient-pole stators for small single-sided disc type AFPM motors fabricated from SMC powders are shown in Fig. 2.11 (*Höganäs*, Sweden) and Fig. 2.12 (*Mii Technologies, LLC*, Lebanon, NH, U.S.A.). The three-phase stator has 9 coils located on salient poles.

2.5 Permanent magnets

2.5.1 Characteristics of PM materials

A PM can produce magnetic flux in an air gap with no field excitation winding and no dissipation of electric power. As any other ferromagnetic material, a PM can be described by its B—H hysteresis loop. PMs are also called *hard magnetic materials*, which means ferromagnetic materials with a wide hysteresis loop. The basis for the evaluation of a PM is the portion of its hysteresis loop located in the upper left-hand quadrant, called the *demagnetization curve*. For rare earth PM at room temperature the demagnetization curve is approximately a straight line drawn between two points: remanent magnetic flux density B_r at $H = 0$ and coercivity H_c at $B = 0$.

Compared to ferrites and Alnico, rare-earth PMs enable a considerable reduction in volume of excitation systems, higher air gap magnetic flux densities, and better dynamic performance of electric motors. Characteristics of representative PM materials are shown in Table 2.12. Nowadays, most PM brushless motors use NdFeB magnets.

Research efforts are oriented towards increasing the magnetic energy density, service temperature, and reducing the cost of fabrication. Examples of super high energy density sintered NdFeB magnets are shown in Table 2.13. The remanent magnetic flux density of modern NdFeB magnets exceeds 1.4 T. It is necessary to admit that the highest energy NdFeB PMs developed so far have low maximum continuous service temperature (approximately 120°C or less for $B_r \geq 1.4$ T).

Since the temperature coefficients of NdFeB are high, NdFeB magnets cannnot be used at temperatures exceeding 200° C. SmCo PMs, although with

Table 2.12. Characteristics of typical PM materials used in electrical machines

Property	Ferrite Ceramic 8	Alnico Alloy	SmCo Sm2Co17	NdFeB Bonded	NdFeB Sintered
B_r, T	0.4	1.25	1.0 to 1.1	0.55 to 0.70	1.25 to 1.35
H_c, kA/m	270	55	600 to 800	180 to 450	950 to 1040
iH_c, kA/m	260	55	720 to 2000	210 to 1100	1200 to 1400
$(BH)max$, kJ/m^3	25 to 32	< 44	190 to 240	32 to 88	290 to 400
α_B, %/°C	−0.20	−0.02	−0.03	−0.105	−0.11
α_H, %/°C	−0.27	−0.015	−0.15	−0.4	−0.65
T_c, °C	460	890	800	360	330

α_B is the reversible temperature coefficient of B_r.
α_H is the reversible temperature coefficient of H_c.
T_c is Curie temperature.

Table 2.13. Examples of super high energy density sintered NdFeB PM materials.

Manufacturer	Grade	B_r, T	H_c, kA/m	iH_c, kA/m	$(BH)max$, kJ/m^3
Hitachi	HS-50AH	1.38 to 1.45	1042 to 1130	≥ 1035	358 to 406
	HS-47AH	1.35 to 1.42	1019 to 1106	≥ 1115	342 to 390
	HS-46CH	1.33 to 1.40	1003 to 1090	≥ 1353	334 to 374
ShinEtsu	N50	1.38 to 1.43	≥ 820	≥ 875	366 to 405
	N48M	1.35 to 1.40	≥ 995	≥ 1114	350 to 390

slightly lower B_r, are characterized by much smaller temperature coefficients and are used in harsh environment, e.g., turbine engine driven generators and oil cooled high speed, high power density machines.

2.5.2 Rare-earth permanent magnets

The first generation of *rare-earth permanent magnets*, i.e., alloys based on the composition of SmCo$_5$ has been commercially produced since the early 1970s (invented in the 1960s). SmCo$_5$/Sm$_2$Co$_{17}$ has the advantage of a high remanent flux density, high coercive force, high energy product, a linear demagnetization curve and a low temperature coefficient (Table 2.14). The temperature coefficient of B_r is −0.02 to −0.045%/°C and the temperature coefficient of H_c is −0.14 to −0.40%/°C. Maximum service temperature is 300 to 350°C. It is suitable for motors with low volumes and motors operating at increased temperatures, e.g. brushless generators for microturbines. Both Sm and Co are relatively expensive due to their supply restrictions.

With the discovery in the late 1970s of a second generation of rare-earth magnets on the basis of inexpensive neodymium (Nd), remarkable progress with regard to lowering raw material costs has been achieved. The new generation of rare-earth PMs based on inexpensive neodymium (Nd) was announced

Table 2.14. Physical properties of Vacomax sintered Sm_2Co_{17} PM materials at room temperature $20°C$ manufactured by *Vacuumschmelze GmbH*, Hanau, Germany

Property	Vacomax 240 HR	Vacomax 225 HR	Vacomax 240
Remanent flux density, B_r, T	1.05 to 1.12	1.03 to 1.10	0.98 to 1.05
Coercivity, H_c, kA/m	600 to 730	720 to 820	580 to 720
Intrinsic coercivity, iH_c, kA/m	640 to 800	1590 to 2070	640 to 800
$(BH)_{max}$, kJ/m³	200 to 240	190 to 225	180 to 210
Relative recoil magnetic permeability	1.22 to 1.39	1.06 to 1.34	1.16 to 1.34
Temperature coefficient α_B of B_r at 20 to 100°C, %/°C		−0.030	
Temperature coefficient α_{iH} of iH_c at 20 to 100°C, %/°C	−0.15	−0.18	−0.15
Temperature coefficient α_B of B_r at 20 to 150°C, %/°C		−0.035	
Temperature coefficient α_{iH} of iH_c at 20 to 150°C, %/°C	−0.16	−0.19	−0.16
Curie temperature, °C		approximately 800	
Maximum continuous service temperature, °C	300	350	300
Thermal conductivity, W/(m °C)		approximately 12	
Specific mass density, ρ_{PM}, kg/m³		8400	
Electric conductivity, ×10⁶ S/m		1.18 to 1.33	
Coefficient of thermal expansion at 20 to 100°C, ×10⁻⁶/°C		10	
Young's modulus, ×10⁶ MPa		0.150	
Bending stress, MPa	—	90 to 150	
Vicker's hardness		approximately 640	

by *Sumitomo Special Metals*, Japan, in 1983 at the 29th Annual Conference of Magnetism and Magnetic Materials held in Pittsburgh, PA, U.S.A. The Nd is a much more abundant rare-earth element than Sm. NdFeB magnets, which are now produced in increasing quantities have better magnetic properties (Table 2.15) than those of SmCo, but unfortunately only at room temperature. The demagnetization curves, especially the coercive force, are strongly temperature dependent. The temperature coefficient of B_r is −0.09 to −0.15%/°C and the temperature coefficient of H_c is −0.40 to −0.80%/°C. The maximum service temperature is 200 to 250°C and Curie temperature is 350°C). The NdFeB is also susceptible to corrosion. NdFeB magnets have great potential for considerably improving the *performance–to–cost* ratio for many applications. For this reason they have a major impact on the development and application of PM machines.

Chemical reactivity of rare-earth magnets is similar to that of alkaline earth metals, e.g. magnesium. The reaction is accelerated at increased temperature and humidity. The NdFeB alloy, if exposed to hydrogen gas, usually at a slightly elevated temperature and/or elevated pressure, becomes brittle and with very little effort, it can be crushed. Diffusion of hydrogen into the alloy causes it literally to fall apart.

Corrosion protective coatings can be divided into metallic and organic. For metallic coatings, e.g., nickel and tin, galvanic processes are used as a rule. Organic coatings include powder coatings applied electrostatically, varnishes and resins.

Table 2.15. Physical properties of Hicorex-Super sintered NdFeB PM materials at room temperature 20°C manufactured by *Hitachi Metals, Ltd.*, Tokyo, Japan

Property	Hicorex-Super HS-38AV	Hicorex-Super HS-25EV	Hicorex-Super HS-47AH
Remanent flux density, B_r, T	1.20 to 1.30	0.98 to 1.08	1.35 to 1.43
Coercivity, H_c, kA/m	875 to 1035	716 to 844	1018 to 1123
Intrinsic coercivity, iH_c, kA/m	min. 1114	min. 1989	min. 1114
$(BH)_{max}$, kJ/m³	278 to 319	183 to 223	342 to 390
Relative recoil magnetic permeability		1.03 to 1.06	
Temperature coefficient α_B of B_r at 20 to 100°C, %/°C		−0.11 to −0.13	
Temperature coefficient α_{iH} of iH_c at 20 to 100°C, %/°C		−0.65 to −0.72	
Curie temperature, °C		≈ 310	
Maximum continuous service temperature, °C	160	180	140
Thermal conductivity, W/(m°C)		≈ 7.7	
Specific mass dens., ρ_{PM}, kg/m³		7500	
Electric conductivity, ×10⁶ S/m		≈ 0.67	
Coefficient of thermal expansion at 20 to 100°C, ×10⁻⁶/°C		−1.5	
Young's modulus, ×10⁶ MPa		0.150	
Bending stress, MPa		260	
Vicker's hardness		≈ 600	
Features	High energy product	High temperature	Super high performance

Nowadays, for the industrial production of rare-earth PMs the powder metallurgical route is mainly used. Apart from some material specific parameters, this processing technology is, in general, the same for all rare-earth magnet materials. The alloys are produced by vacuum induction melting or by a calciothermic reduction of the oxides. The material is then size-reduced by crushing and milling to a single crystalline powder with particle sizes less than 10 μm. In order to obtain anisotropic PMs with the highest possible $(BH)_{max}$ value, the powders are then aligned in an external magnetic field, pressed and densified to nearly theoretical density by sintering. The most economical method for mass production of simply shaped parts like blocks, rings or arc segments is *die pressing* of the powders in approximately the final shape.

Researchers at *General Motors*, MI, U.S.A. developed a fabrication method based on the melt-spinning casting system originally invented for the production of amorphous metal alloys. In this technology a molten stream of NdFeCoB material is first formed into ribbons 30 to 50-μm thick by rapid quenching, then cold pressed, extruded and hot pressed into bulk. Hot pressing and hot working are carried out while maintaining the fine grain to provide a high density close to 100% which eliminates the possibility of internal corrosion. The standard electro-deposited epoxy resin coating provides excellent corrosion resistance.

The prices of NdFeB magnets ordered in large quantities are now around US$50 per kg. Owing to a large supply of NdFeB magnets from China it is expected that the prices will fall further.

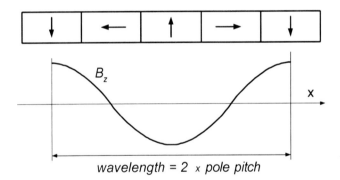

Fig. 2.13. Cartesian Halbach array. B_z is the normal component of the magnetic flux density.

2.5.3 Halbach array

Cylindrical rotor PMBMs and twin rotors of double-sided coreless AFPM machines (Section 5.5) may use PMs arranged in *Halbach array* [78, 79, 80]. The key concept of Halbach array is that the magnetization vector of PMs should rotate as a function of distance along the array (Fig. 2.13) [78, 79, 80]. Halbach array offers the following advantages:

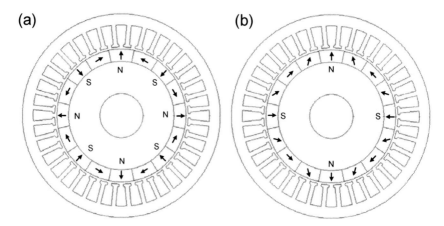

Fig. 2.14. PMBMs with Halbach magnetized rotors: (a) 8-pole rotor with 90^0 Halbach array; (b) 4-pole motor with 45^0 Halbach array. Images have been produced with the aid of SPEED software, University of Glasgow, U.K.

- the fundamental field is stronger by 1.4 than in a conventional PM array, and thus the power efficiency of the machine is approximately doubled;
- the array of PMs does not require any backing steel magnetic circuit and PMs can be bonded directly to a non-ferromagnetic supporting structure (aluminum, plastics);
- the magnetic field is more sinusoidal than that of a conventional PM array;
- non-overlapping concentrated coil stator winding can be used (Section 3.3);
- Halbach array has very low back-side fields.

Fig. 2.14a shows a rotor of 8-pole PMBM with 90^0 Halbach array and Fig. 2.14b shows a 4-pole PMBM with 45^0 Halbach array.

2.6 Wire insulation with heat activated adhesives

Recently, a new method of conductor securing that does not require any additional material, and uses very low energy input, has emerged [115]. The solid conductor wire (usually copper) is coated with a *heat and/or solvent activated adhesive*. The adhesive which is usually a polyvinyl butyral, utilizes a low temperature thermoplastic resin [115]. This means that the bonded adhesive can come apart after a certain minimum temperature is reached, or it again comes in contact with the solvent. Normally, this temperature is much lower than the thermal rating of the base insulation layer. The adhesive is activated by either passing the wire through a solvent while winding or heating the finished coil as a result of passing electric current through it.

The conductor wire with a heat activated adhesive overcoat costs more than the same class of non-bondable conductor. However, a less than two second current pulse is required to bond the heat activated adhesive layer and bonding machinery costs about half as much as trickle impregnation machinery [115].

2.7 High temperature materials

2.7.1 High temperature ferromagnetic materials

Aircraft integrated starter/generators (ISGs) mounted on the central shaft of the gas turbine engine or electromechanical actuators for more electric engines (MEE) may require operating temperatures at least 600^0C [61]. Co-V-Fe alloys (up to 50% cobalt) can provide saturation magnetic flux density of approximately 1.6 T at 850^0C. Hiperco 50HS, Hiperco 50 and Hiperco 27 from *Carpenter*, U.S.A., and AFK1 from *Imphy s.a.*, France, are currently used for prototypes of high temperature reluctance machines and short stroke actuators.

2.7.2 High temperature insulating materials and conductors

Table 2.16. Maximum temperature rise $\Delta \vartheta$ for armature windings of electrical machines according to IEC and NEMA (based on $40°$ ambient temperature)

Rated power of machines, length of core and voltage	Insulation class				
	A °C	E °C	B °C	F °C	H °C
IEC a.c. machines < 5000 kVA (resistance method)	60	75	80	100	125
IEC a.c. machines ≥ 5000 kVA or length of core ≥ 1 m (embedded detector method)	60	70	80	100	125
NEMA a.c. machines ≤ 1500 hp (embedded detector method)	70	–	90	115	140
NEMA a.c. machines > 1500 hp and ≤ 7 kV (embedded detector method)	65	–	85	110	135

The *maximum temperature rise* for the windings of electrical machines is determined by the temperature limits of insulating materials. The maximum temperature rise in Table 2.16 assumes that the temperature of the cooling medium $\vartheta_c \leq 40°$C. The maximum temperature of windings is actually $\vartheta_{max} = \vartheta_c + \Delta\vartheta$ where $\Delta\vartheta$ is the maximum allowable temperature rise according to Table 2.16.

Polyester, epoxy or silicon resins are used most often as impregnating materials for treatment of stator windings. Silicon resins of high thermal endurance are able to withstand $\vartheta_{max} > 225°$C.

Heat-sealable Kapton® polyimide films are used as primary insulation on magnet wire at temperatures 220 to 240°C [54]. These films are coated with or laminated to Teflon® fluorinated ethylene propylene (FEP) fluoropolymer, which acts as a high temperature adhesive. The film is applied in tape form by helically wrapping it over and heat-sealing it to the conductor and to itself.

Windings that operate at temperatures over 600°C (aerospace, nuclear, plasma physics, steam and chemical applications) require *nickel clad copper*, *chrome-iron clad copper*, *all nickel* or *palladium-silver* conductor wires with ceramic insulation. Ceramic coating is primarily a refractory-glass compound with flexibility and braiding properties consistent with normal coil practices. Ceramic insulations are designed to be potted after they are wound into components. The coating thickness is from 0.008 to 0.015 mm, depending on size [39]. If the packing (filling) factor is less important, wires with ceramic insulation can be replaced by fiberglass sleeved wires.

2.8 Superconductors

The first commercial low temperature superconducting (LTS) wire was developed at *Westinghouse* in 1962. Typical LTS wires are: magnesium diboride MgB_2 tapes (cost < 10 \$/kAm at 20K and 2 T), NbTi Standard (cost 4 to 6 \$/kAm at 4K and 8 T) and Nb_3Sn Standard (cost 15 to 30 \$/kAm at 4K and 12 T and 75 to 150 \$/kAm at 2K and 21 T). LTS wires are still preferred in high field magnets for nuclear magnetic resonance (NMR), magnetic resonance imaging (MRI), magnets for accelerators and fusion magnets. Input cooling power as a function of required temperature is shown in Fig. 2.15.

Manufacturers of electrical machines need low cost HTS tapes that can operate at temperatures approaching 77 K for economical generators, motors and other power devices. The minimum length of a single piece acceptable by electrical engineering industry is at least 100 m. The U.S. Department of Energy (DoE) target cost is \$50/kA-m in 2007, \$20/kA-m in 2009 and \$10/kA-m after 2010. It is rather unlikely that this cost target will be met.

2.8.1 Classification of HTS wires

In 1986 a family of cuprate-perovskite ceramic materials known as high temperature superconductors (HTS) with critical temperatures in excess of 77K

2.8 Superconductors   51

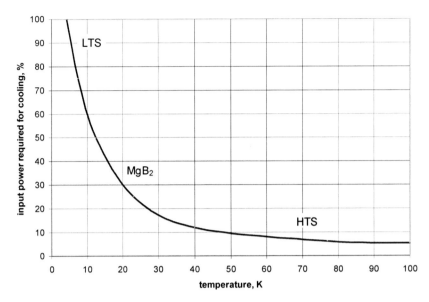

Fig. 2.15. Input cooling power in percent of requirements at 4.2K versus temperature.

was discovered. HTS refers to materials with much higher transition temperatures (liquid nitrogen, 77K or -196°C), than previously known low temperature superconductors (LTS) cooled with liquid helium (4.2K or about -269°C transition temperature).

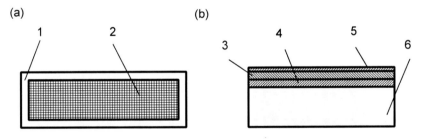

Fig. 2.16. HTS wires: (a) 1st generation (1G); (b) 2nd generation (2G). 1 — silver alloy matrix, 2 — SC filaments, 3 — SC coating, 4 — buffer layer, 5 — noble metal layer, 6 — alloy substrate.

HTS wires are divided into three categories:

- *First generation* (1G) superconductors, i.e., *multi-filamentary tape conductors* BiSrCaCuO (BSCCO) developed up to industrial state. Their properties are reasonable for different use, but prices are still high.

- *Second generation* (2G) superconductors, i.e., *coated tape conductors*: YBaCuO (YBCO) which offer superior properties. They are commercially manufactured.
- *Third generation* (3G) superconductors with *enhanced pinning*[2] (early study).

Fig. 2.16 shows basic 1G and 2G HTS wire tape architecture according to *American Superconductors* [9]. Each type of advanced wire achieves high power density with minimal electrical resistance, but differs in the superconductor materials manufacturing technology, and, in some instances, its end-use applications.

Parameters characterizing superconductors are:

- critical current I_c × (wire length), Am (200 580 Am in 2008) [188];
- critical current I_c/(wire width), A/cm–width (700 A/cm–width in 2009) [9];
- critical current density J_c, A/cm^2;
- engineering critical current density J_e, A/cm^2.

The *engineering critical current density* J_e is the critical current of the wire divided by the cross sectional area of the entire wire, including both superconductor and other metal materials.

BSCCO 2223 is a commonly used name to represent the HTS material $Bi_{(2-x)}Pb_xSr_2Ca_2Cu_3O_{10}$. This material is used in multi-filamentary composite HTS wire and has a typical superconducting transition temperature around 110 K. BSCCO-2223 is proving successful presently, but will not meet all industrial requirements in the nearest future. According to *SuperPower* [188], there are clear advantages to switch from 1G to 2G, i.e.,

- better in-field performance;
- better mechanical properties (higher critical tensile stress, higher bend strain, higher tensile strain);
- better uniformity, consistency, and material homogeneity;
- higher engineering current density;
- lower a.c. losses.

There are key areas where 2G needs to be competitive with 1G in order to be used in the next round of various device prototype projects. Key benchmarks to be addressed are :

- long piece lengths;
- critical current over long lengths;

---

[2] Magnetic flux lines do not move (become trapped, or *pinned*) in spite of the Lorentz force acting on them inside a current-carrying type II SC. The phenomenon cannot occur in type I SC, since these cannot be penetrated by magnetic fields (Meissner-Ochsenfeld effect).

- availability (high throughput, i.e., production volume per year, large deliveries from pilot-scale production);
- comparable cost with 1G.

Power applications of HTS wires include:

- transformers;
- power cables;
- large coils, e.g., nuclear magnetic resonance (NMR), magnetic separations, accelerator electromagnets;
- fault current limiters (FCL);
- UPS backups – SC magnet energy storage system(SMES);
- active power filters;
- electric motors and generators;
- generation of strong magnetic fields;
- medical applications, e.g., MRI.

Commercial quantities of HTS wire based on BSCCO are now available at around five times the price of the equivalent copper conductor. Manufacturers are claiming the potential to reduce the price of YBCO to 50% or even 20% of BSCCO. If the latter occurs, HTS wire will be competitive with copper in many large industrial applications.

Recently discovered magnesium diboride MgB_2 is a much cheaper SC than BSCCO and YBCO in terms of dollars per current carrying capacity times length (\$/kA-m). Most manufactured wires are already substantially cheaper than copper. This is partly because the price of copper has recently increased. However, this material must be operated at temperatures below 39K, so the cost of cryogenic equipment is very significant. This can be half the capital cost of the machine when cost effective SC is used. Magnesium diboride SC MgB_2 might gain niche applications, if further developments will be successful.

As of March 2007, the current world record of superconductivity is held by a ceramic SC consisting of thallium, mercury, copper, barium, calcium, strontium and oxygen with $T_c = 138$ K.

2.8.2 HTS wires manufactured by American Superconductors

American Superconductors (AMSC) [9] is the world's leading developer and manufacturer of HTS wires. Fig. 2.16a schematically shows the internal structure of the 1G multi-filamentary composite HTS wire. Fig. 2.16b shows the basic architecture for 2G SC wire, which the company has branded as 344 and 348 superconductors. 2G 344 superconductors were introduced to the market in 2005. AMSC HTS wire specifications are given in Table 2.17.

Today, AMSC 2G HTS wire manufacturing technology is based on 100 meter-long, 4-cm wide strips of SC material (Fig. 2.17) that are produced

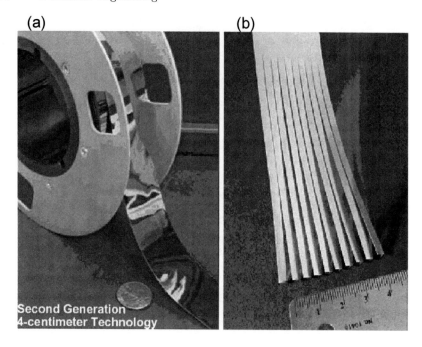

Fig. 2.17. Second generation (2G) 4 cm technology: (a) 4 cm wide strip; (b) strip cut into 4 mm tapes. Photo courtesy *American Superconductors* [9].

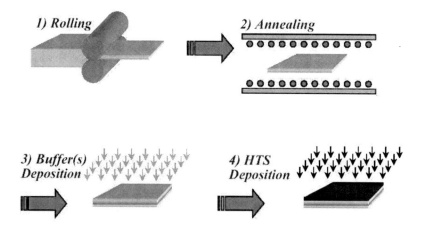

Fig. 2.18. High-speed, continuous reel-to-reel deposition process RABITS. Courtesy of *American Superconductors* [9].

2.8 Superconductors

Table 2.17. Specifications of HTS wires manufactured by *American Superconductors*, Westborough, MA, U.S.A. [9].

Specifications	Bismuth based, multi filamentary HTS wire encased in a silver alloy matrix	344 HTS copper stabilized wires, 4.4 mm wide	344 HTS stainless steel stabilized wires, 4.4 mm wide
Grade	BSCC0, 1G	YBCO, 2G	YBCO, 2G
Average thickness, mm	0.21 to 0.23	0.20 ± 0.02	0.15 ± 0.02
Minimum width, mm	3.9		
Maximum width, mm	4.3	4.35 ± 0.05 average	4.33 ± 0.07 average
Minimum double bend diameter at 20oC, mm	100	30	30
Maximum rated tensile stress at 20°C, MPa	65	150	150
Maximum rated wire tension at 20°C, kg	4		
Maximum rated tensile stress at 77K, MPa	65		
Maximum rated tensile strain at 77K, %	0.10	0.3	0.3
Average engineering current density J_e, A/cm^2, at minimum critical current I_c, A	$J_e = 12,700\ I_c = 115$ $J_e = 13,900\ I_c = 125$ $J_e = 15,000\ I_c = 135$ $J_e = 16,100\ I_c = 145$	$J_e = 8000$ $I_c = 70$	$J_e = 9200$ $I_c = 60$
Continuous piece length, m	up to 800	up to 100	up to 20

in a high-speed, continuous reel–to–reel deposition [3] process, the so–called *rolling assisted biaxially textured substrates* (RABITS). RABITS is a method for creating textured[4] metal substrate for 2G wires by adding a buffer layer between the nickel substrate and YBCO. This is done to prevent the texture of the YBCO from being destroyed during processing under oxidizing atmospheres (Fig. 2.18). This process is similar to the low-cost production of motion picture film in which celluloid strips are coated with a liquid emulsion and subsequently slit and laminated into eight, industry-standard 0.44-cm-wide tape-shaped wires (344 superconductors). The wires are laminated on both sides with copper or stainless-steel metals to provide strength, durability and certain electrical characteristics needed in applications. AMSC RABiTS/MOD (metal organic deposition) 2G strip architecture is as follows (Fig.

---

[3] Deposition (in chemistry) is the settling of particles (atoms or molecules) or sediment from a solution, suspension mixture or vapor onto a pre-existing surface.

[4] Texture in materials science is the distribution of crystallographic orientations of a sample. Biaxially textured means textured along two axes.

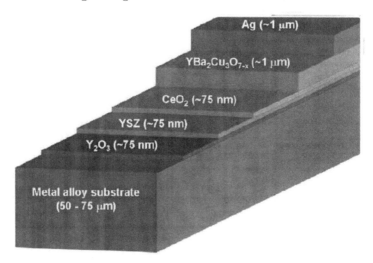

Fig. 2.19. RABITS/MOD 2G HTS architecture. Courtesy of *American Superconductors* [9].

2.19): (1) substrate: Ni-5%W alloy (deformation texturing), (2) buffer stack: $Y_2O_3/YSZ/CeO_2$ (high rate reactive sputtering[5]), (3) YBCO (metal organic deposition of trifluoroacetate (TFA) based precursors), (4) Ag (sputtering). AMSC expects to scale up the 4 cm technology to 1000 meter lengths. The company then plans to migrate to 10-cm technology to further reduce manufacturing costs.

Sumitomo Electric Industries, Japan, uses holmium (Ho) rare element instead of yttrium (Y). According to *Sumitomo*, the HoBCO SC layer allows for higher rate of deposition, high critical current density J_c and better flexibility of tape than the YBCO SC layer.

2.8.3 HTS wires manufactured by SuperPower

SuperPower [188] uses *ion beam assisted deposition* (IBAD), a technique for depositing thin SC films. IBAD combines ion implantation with simultaneous sputtering or another physical vapor deposition (PVD) technique. An ion beam is directed at an angle towards the substrate to grow textured buffer layers. According to *SuperPower*, virtually, any substrate could be used, i.e., high-strength substrates, non-magnetic substrates, low cost, off-the shelf substrates (Inconel, Hastelloy, stainless steel), very thin substrates, resistive substrates (for low a.c. losses), etc. There are no issues with percolation (trickling or filtering through a permeable substance). IBAD can pattern conductor to

---

[5] Sputter deposition is a physical vapor deposition (PVD) method of depositing thin films by sputtering, i.e., eroding, material from a 'target', i.e., source, which then deposits onto a "substrate," e.g., a silicon wafer.

very narrow filaments for low a.c. loss conductor. An important advantage is small grain size in sub micron range.

IBAD MgO based coated conductor has five thin oxide buffer layers with different functions, i.e.,

1. alumina barrier layer to prevent diffusion of metal element into SC;
2. yttria seed layer to provide good nucleation surface for IBAD MgO;
3. IBAD MgO template layer to introduce biaxial texture;
4. homo-epitaxial MgO buffer layer to improve biaxial texture;
5. SrTiO$_3$ (STO) cap layer, to provide lattice match between MgO and YBCO and good chemical compatibility.

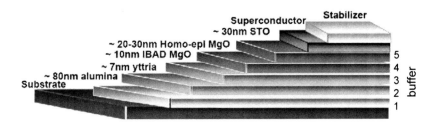

Fig. 2.20. IBAD MgO based coated conductor. The buffer consists of five oxide layers: 1 — alumina, 2 — yttria, 3 — IBAD MgO, 4 — homo-epitaxial MgO, 5 — SrTiO$_3$ (STO). Courtesy of *SuperPower*, Schenectady, NY, U.S.A. [188].

Tables 2.18 and 2.19 show the cost and selected specifications of HTS 2G tapes manufactured by *SuperPower* [188].

Table 2.18. Cost per meter and parameters of SCS4050 4mm HTS tape [188].

SCS4050	$/m	I_c at 4 mm width, 77 K, self field, A	$/kA-m	width with copper stabilizer, mm
2006	100	80	1250	4
2007	65	100	650	4

Table 2.19. Cost per meter and parameters of SF12050 12mm HTS tape [188].

SF12050	$/m	I_c at 12 mm width, 77 K, self field, A	$/kA-m	width without copper stabilizer, mm
2006	150	240	625	12
2007	90	300	300	12

2.8.4 Bulk superconductors

HTS ceramic materials are brittle. Long length HTS wires are manufactured as multifilamentary tapes (1G) or coated tapes (2G). *Bulk HTS monoliths* can be fabricated in large sizes as either single crystal or multiple crystals with flow of currents across the grains. Very good performance of YBCO at 77K is the main reason of using this material for SC magnets, levitators or magnetic bearings (Fig. 2.21). A bulk HTS is a monolith without leads. It is inductively charged with the aid of external magnetic fields. Inexpensive commercial HTS monoliths with size of minimum 25 mm are now available on the market (Fig. 2.22).

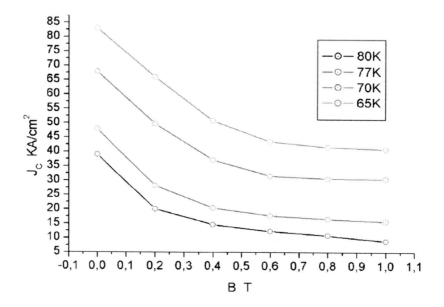

Fig. 2.21. Critical current at different temperatures and magnetic fields for bulk YBCO HTS. Courtesy of *Nexans*, Paris, France [137]

A quadruple magnet made of bulk HTS discs is shown in Fig. 2.23 [33]. Such a magnet can be assembled at room temperature and then charged after assembly. The distribution of forces is similar to that of multiple PM system. HTS monoliths behave similar to PMs with very low Curie temperature, i.e., critical temperature of HTS. This magnetic flux density is much higher than 1 T which is approximately maximum flux density obtained using the highest energy density PMs. In a symmetrical quadruple magnet the forces on HTS discs are in radial direction, i.e., there is no net force.

One of the techniques of inductive charging of bulk HTS monoliths is called *flux trapping*. A significant magnetic field can be trapped by a superconductor

2.8 Superconductors 59

Fig. 2.22. Large grain YBCO bulk monoliths.

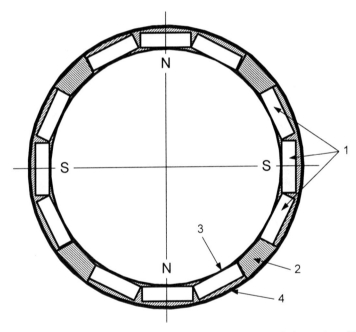

Fig. 2.23. Quadrupole magnet assembled of twelve bulk HTS discs: 1 — HTS disc (monolith), 2 — supporting wedges, 3 — cold bore, 4 — external supporting band [33].

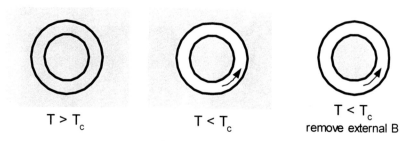

Fig. 2.24. Magnetic field trapped in a superconducting cylinder.

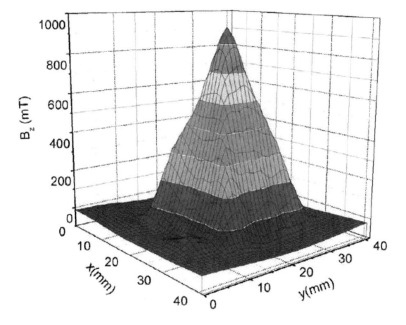

Fig. 2.25. Test results of magnetic field trapped in an YBCO bulk superconducting material. Courtesy of *Nexans*, Paris, France [137].

when it exhibits large flux pinning forces showing that the material becomes a quasi PM.

If an SC ring or hollow cylinder is placed in the magnetic field, temperature is reduced below the critical temperature T_c and then, the external magnetic field is removed, the magnetic field will be trapped in an SC ring or cylinder (Fig. 2.24). The field freezing method applies magnetic field to bulk HTS material at room temperature. Then, the temperature is decreased until the monolith becomes superconducting and the external magnetic field is removed.

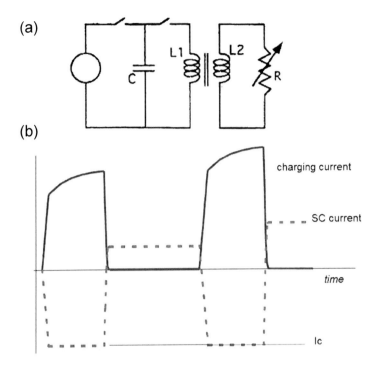

Fig. 2.26. Inductive charging for bulk HTS monoliths: (a) charging circuit; (b) current waveforms in primary and secondary coils [33].

Magnetic flux density distribution trapped at the surface of a YBCO sample is shown in Fig. 2.25 [137]. Using small single crystal YBCO discs magnetic flux densities higher than 10 T can be obtained at 77 K.

A HTS can be brought to its critical state at near-constant temperature by applying additional fields or currents [33]. When in critical state, the magnetic field of the HTS can freely move in and out (flux pumping). There is some dissipation of the magnetic field because the temperature slightly increases as a result of large thermal capacity at 20 to 50 K. There is no quench[6] and little evidence of flux jumping at this temperature [33].

The charging circuit is shown in Fig. 2.26a [33]. The capacitor C is charged by IGBT or MOSFET-based power supply and discharged through the charging (primary) coil L_1. The HTS is represented by the inductance L_2 and variable resistance R. When HTS is beyond its critical state, the resistance $R = 0$. The current waveforms in the charging and secondary coils for an HTS monolith are shown in Fig. 2.26b [33]. The process of charging is repeated until the required current in the bulk HTS is reached.

---

[6] Quench is a term commonly used to describe the process which occurs when any part of an SC coil goes from the SC to the normal resistive state.

2.9 Nanostructured materials

Nanostructured materials are constituted of building blocks of metals, ceramics or polymers that are nanometer–size objects[7]. They can be defined as materials the structural elements of which, i.e., clusters, crystallites or molecules have dimensions in the 1 to 100 nm range. Typical nanostructured materials are thin films (nanolitography), nanotubes, nanocrystalline materials, nano composites, biologically-based nanostructures. Physical properties of the above materials are novel and can be directed by controlling the dimensions of building blocks and their assembly via physical, chemical or biological methods. For example, the strength of pure metals is increased when grain sizes are reduced to below 50 nm, strength and electric conductivity of carbon nanotubes is very high or ductility in nanostructured ceramics is enhanced.

2.9.1 Carbon nanotubes

Carbon nanotubes were discovered in 1991, by Japanese researcher S. Iijima. A *carbon nanotube* (CNT) is a hexagonal network of carbon atoms rolled to form a seamless cylinder (Fig. 2.27a). There are two types of CNTs: a *single wall carbon nanotube* (SWNT) which consists of one cylinder (Fig. 2.27a) and *multi-wall nanotube* (MWNT) composed of several concentric graphene cylinders (Fig. 2.27a). CNTs can carry enormous current density 10^5 to 10^7 A/mm^2 before failing as a result of self-electrolysis. Their Young modulus is 1.0 to 5.0 TPa for SWNT and about 1 TPa for MWNT in comparison with 200 GPa for steel. The CNT tensile strength is 10 to 150 GPa versus only 0.4 GPa for steel. Table 2.20 compares dimensions of nanotubes with dimensions of other objects from hydrogen atom to leukocyte.

Suggested applications for CNTs include nano-electronic devices, transistors, field effect displays, sensors, nano-electromechanical machines, energy storage, and even a *space elevator*. This last, science fiction-like, device is theoretically possible due CNTs very high tensile strength. Super-strong CNTs may make space elevators feasible. No other material can withstand its own weight for such a height. Some predicted specifications of a space elevator are:

- ribbon tension = 25 t (force);
- climber total mass = 20 t;
- fixed mass overhead = 5 t;
- mass–to–power = 1 t/MW;
- feed speed limit = 100 m/s = 360 km/h;
- climber power capacity = 5 MW.

If a counterweight is attached to a cable, and put far enough away, about 100 000 km, the cable will be held taut by the force of the Earth's rotation,

---

[7] Greek word *nanos* means *dwarf*.

2.9 Nanostructured materials 63

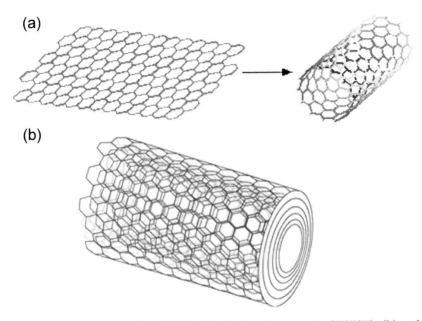

Fig. 2.27. Carbon nanotubes (CNTs): (a) single wall nanotube (SWNT); (b) multi-wall nanotube (MWNT).

Table 2.20. Dimensions of objects in nanometers

Hydrogen atom	0.1
Fulleren C_{60}	1.0
Six connected atoms of carbon	1.0
DNA	2.0
Quantum dot	2 to 10
Nanoparticle	2 to 100
Nanotube	3 to 30
Proteins	5 to 50
Dip-pen nanolitography	10 to 15 nm
Dendrimer	10 to 20 nm
Microtubula	25
Ribosome	25
Semiconductor chip	65 and more
Virus	75 to 100
Mitochondrium	500 to 1000
Bacteria	1000 to 10,000
Red blood cell (erythrocyte)	6000 to 8000
Capillary	8000
White blood cell (leukocyte)	10,000

64   2 Material engineering

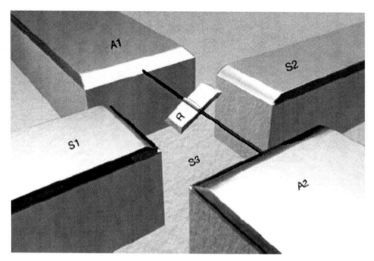

Fig. 2.28. Piezoelectric actuator fabricated in the Department of Physics at the University of California at Berkeley, CA, U.S.A. R — rotor metal plate, A_1 and A_2 — electrically conducting anchors, S_1, S_2 and S_3 (bottom) — electrodes (Si substrate). MWNT between A_1 and A_2 provides support shaft, source of rotational freedom and electrical feed [59].

similar to spinning around a ball on a string, while holding it. A taut cable is a track for an elevator.

There are two concepts: an elevator which can reach the *geosynchronous circular orbit* (about 35 800 km from the earth equator surface) and elevator which can reach 100 000 km into space. The space elevator starts with a basic floating platform in the ocean, near the equator. Attached to the platform is a paper-thin ribbon not more than 1 m wide that stretches 100 000 kilometres into space, about one-quarter of the way to the moon. There it is tied to a satellite that pulls the ribbon taut and keeps it straight as it orbits in synchronism with the Earth's rotation. An elevator car could climb and descend the ribbon at a speed up to 360 km/h. That means the trip to geosynchronous orbit (35 800 km) would last over four days.

The 100 000-km long space elevator would rise far above the average orbiting height of the space shuttle (185 to 643 km). This new space transportation system could make travel to *geostationary orbit* (zero inclination geosynchronous orbit) a daily event and transform the global economy.

CNTs can theoretically be used in various power applications, such as cables, electromechanical generators, motors, electromagnets and transformers. However, the length of currently available single fiber CNTs is severely restricted which excludes any power engineering application. A very small, about 300 nm piezoelectric actuator (Fig. 2.28) has yet been fabricated, that within the Department of Physics at the University of California at Berkeley [59]. The rotational solid rectangular metal plate R (rotor) is attached transversely to the suspended support shaft. An MWNT serves simultaneously as the rotor plate support shaft, feeds the rotor plate with current and provides rotational freedom. The support shaft ends are embedded in electrically conducting anchors A_1 and A_2 that rest on the oxidized surface of a silicon chip. The rotor plate assembly is surrounded by three stationary electrodes: S_1 and S_2 at both sides of the rotor and the third *gate* electrode S_3 is placed below the rotor. Four independent (d.c. and/or a.c.) voltage signals (one to the rotor plate and three to the stationary electrodes) can be applied to control the position, speed and direction of rotation of the rotor plate.

In 2004, scientists at Henry Samueli School of Engineering, University of California Irvine, U.S.A. synthesized a 4-mm electrically conducting nanotubes. Also in 2004, researchers at Los Alamos National Laboratory, Los Alamos, NM, U.S.A and Duke University, Durham, NC, U.S.A. using catalytic chemical vapor deposition from ethanol created 40-mm long SWNT.

2.9.2 Soft magnetic nanocrystalline composites

Acording to U.S. Navy, major limitation to realizing the potential of large current PWM motor drives and power converters is the lack of efficient and compact switching inductors.

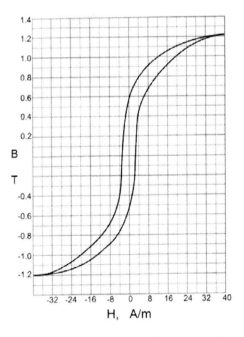

Fig. 2.29. D.c. hysteresis loop of a nanocrystalline composite with high magnetic permeability and low core losses [127].

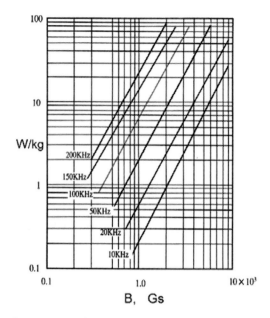

Fig. 2.30. Core loss curves of a nanocrystalline composite with high magnetic permeability and low core losses [127].

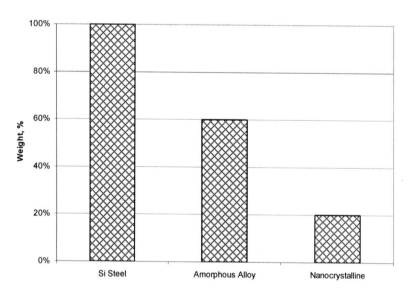

Fig. 2.31. Potential core weight reduction when using nanocrystalline composites and operating at 0.2 T and 20 kHz in comparison with silicon steel and amorphous alloys.

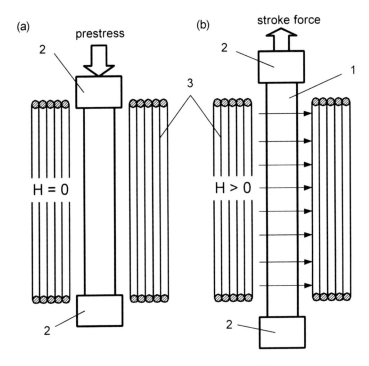

Fig. 2.32. Principle of operation of a MSM actuator: (a) transverse magnetic field $H = 0$; (b) transverse magnetic field $H > 0$. 1 — MSM bar, 2 — PM, 3 — coils.

Nanocrystalline soft magnetic material are new composites of 82% Fe with the remaining balance of silicon (Si), boron (B), niobium (Nb), copper (Cu), carbon (C), molybdenum (Mo) and nickel (Ni). The nanocrystalline grain size is approximately 10 nm. The raw material is manufactured and supplied in an amorphous state. It is recrystalized into a precise mix of amorphous and nanocrystalline phases when annealed, giving the material its unique magnetic properties. The saturation magnetic flux density of nanocrystalline composites approaches 1.2 T, core losses at 0.5 T and 20 kHz are less than 25 W/kg and Curie temperature is approximately 570⁰C. The magnetization curve B–H and core loss curves at different frequencies of nanocrystalline composites with high magnetic permeability and low core losses are shown in Figs. 2.29 and 2.30 [127].

Potential applications are power electronics converters, switching mode power supplies, inverters (motor drives) and high frequency transformers [131]. The power range is 50 kW to 5 MW. Utilizing emerging nanocrystalline composite properties, a nano-inductor reduces the power loss by a factor of 3. It also reduces size and weight by 60 to 80% (Fig. 2.31) and temperature rise by at least 65%.

2.10 Magnetic shape memory materials

Magnetic shape memory (MSM) effect is a new invention in actuator materials field, allowing even 50 times greater strains than in magnetostrictive materials [205]. MSM mechanism essentially differs from magnetostriction. In MSM materials the magnetic field moves microscopic parts of the material (so called twins) that leads to a net shape change of the material. The mechanism enables also more complicated shape changes than conventional linear strain, such as bending and shear.

Table 2.21. Comparison of properties of MSM with other smart actuator materials according to *AdaptaMat*, Helsinki, Findland [150].

	Bulk piezo (PZT)	Multilayered piezo	Magnetostrictive Terfenol-T	MSM Ni$_2$MnGa
Control Field	Electric	Electric	Magnetic	Magnetic
Max. linear strain ϵ, μm/mm,	0.3	1.25	1.6	100
Work output $\sigma_{bl} \times \epsilon$, MPa $\times$ μm/mm)	6	25	112	300
Young's modulus, GPa	48–74	45–62	25–35	7.7
Tensile strength, MPa	5–50	5–30	28	–
Compressive strength, MPa	60	50	700	700
Curie temperature, $^{\circ}$C	200–350	200–350	380	103
Max. operating temperature $^{\circ}$C	100	100	150	70
Resistivity, Ωm	1010	1010	58×10^{-8}	80×10^{-8}
Relative permittivity	800–2400	800–2400	NA	NA
Relative permeability	1	1	3–10	1.5–40
Coupling factor %	75	70	75	75
Max. energy density, kJ/m^3	2	18.5	27	90
Field strength for max. strain	2 MV/m	2 MV/m	240 kA/m	400 kA/m

MSM materials are ferromagnetic alloys in the martensitic state. Under the influence of a magnetic field, these alloys undergo a reorientation of magnetic domains, resulting in large strains. Typical MSM materials used in actuator technology, such as Ni$_2$MnGa produce 2% strain at 0 to 2 MPa stress. Other MSM materials are FePd and FeNiCoTi alloys. The maximum strain of MSM material is about 6%, the shape change occurs in 0.2 ms and the operating temperature range is from –130 to +100°C. Propreties of Ni$_2$MnGa MSM material and its comparison with piezo and magnetostrictive materials in given in Table 2.21. The so–called *blocking stress* σ_{bl} is defined in MSM as the stress at which the strain is 0.01ϵ, where ϵ is maximum strain [150].

Principle of operation of an MSM actuator is explained in Fig. 2.32. The transverse magnetic field is produced by two opposite coils connected in series. Two bias PMs are placed at each end of the sample of Ni_2MnGa material. When the electric current is applied to the coils, the MSM bar elongates in the transverse magnetic field.

MSM materials find numerous applications [150, 205], e.g., in valves, injectors, manipultors, robot grippers , linear drives , switches, shakers, loudspeakers, brakes, clamps, force/position sensors, vibration monitoring, etc.

3
High power density machines

In the last three decades of the 20th century, the following types of high power density electrical machines have been developed:

- PM transverse flux motor (TFM);
- PM brushless disc type motors;
- PM brushless machines with non-overlapping concentrated coils;
- motors for refrigeration compressors;
- motors with cryogenic cooling system.

3.1 Permanent magnet transverse flux motors

The electromagnetic power density of an a.c. machine is proportional to the magnetic (air gap flux density B_g) and electric (stator line current density A) loadings. At constant armature current and diameter, the line current density A of a TFM increases with the number of poles [68]. As it is known, the higher the frequency, the higher the power density of an electrical machine. Since the number of pole pairs is large ($p = 24$ to 144 and sometimes more), the TFM is an excellent machine for direct conversion of high input frequency into low speed or vice versa.

Recent interest in TFMs is due to larger shear stresses $\sigma_p = B_g A$ than in classical longitudinal flux motors. The TFM can be designed as a single sided (Fig. 3.1a) or double sided machine (Fig. 3.1b). Single-sided machines are easier to manufacture and have better prospects for practical applications. TFMs have a simple winding consisting of one ring–shaped coil per phase. A computer generated image of a single phase unit of a TFM with inner stator and outer rotor is shown in Fig. 3.2. A single coil winding is embraced by U–shaped ferromagnetic cores. The rotor consists of large number of PMs. In the case of surface configuration of PMs, the number of PMs per phase is equal to the number of poles $2p$. The magnetic flux in perpendicular (transverse) to

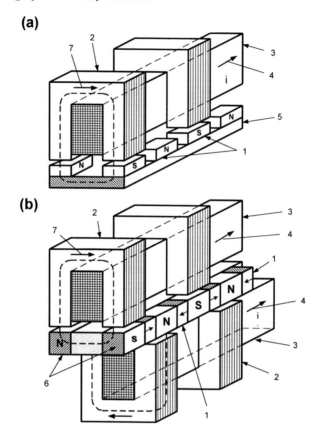

Fig. 3.1. PM transverse flux motor (spread flat): (a) single sided, (b) double sided. 1 — PM, 2 — stator core, 3 — stator winding, 4 — stator current, 5 — rotor yoke, 6 — mild steel poles shoes, 7 — magnetic flux.

the armature current and direction of the electromagnetic torque. Polyphase motors are assembled of single-phase modules (Fig. 3.3).

Fig. 3.4 shows a three-phase, single sided TFM rated at 7.5 kW and 600 rpm. The machine is naturally air cooled without any fan. The number of pole pairs $p = 18$, the air gap is 1 mm and the average air gap diameter is 207 mm.

A double sided, 150 kW, water-cooled TFM for hybrid buses is shown in Fig. 3.5. The maximum speed is 2400 rpm, rated torque 1800 Nm, breakdown torque 2750 Nm and power density 0.8 kW/kg. The water cooling system is located in aluminum parts below the winding. There are two stators: inner and outer stator. The drum-shaped rotor with PMs rotates between the poles of outer and inner stator.

3.1 Permanent magnet transverse flux motors   73

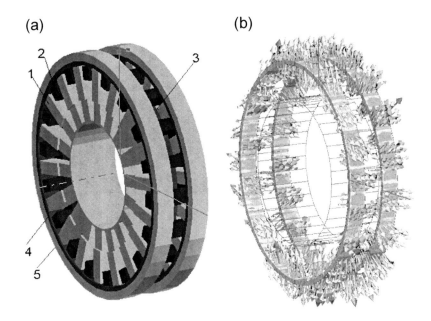

Fig. 3.2. Computer generated image of a single-phase TFM with $p = 24$: (a) stator; (b) magnetic field excited by rotor PMs. Author's simulation.

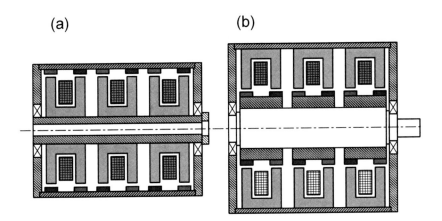

Fig. 3.3. Three phase TFMs: (a) with inner stator; (b) with outer stator.

TFMs have several advantages over standard PM brushless motors, i.e., [68], i.e.,

(a) better utilization of active materials than in standard (longitudinal flux) PM brushless motors for the same cooling system, i.e., higher torque density or higher power density;

Fig. 3.4. Three-phase TFM with inner stator and outer rotor: (a) stator; (b) rotor [71].

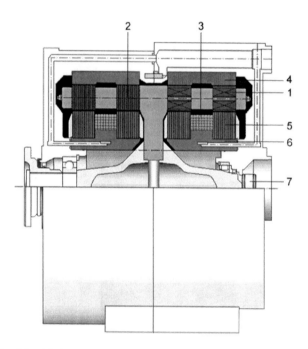

Fig. 3.5. Double-sided 150-kW TFM for hybrid buses. 1 — PM, 2 — rotor lamination pack, 3 — coil, 4 — soft ferromagnetic element, 5 — stator pack of laminations, 6 — soft ferromagnetic yoke, 7 — resolver. Courtesy of *Voith Turbo GmbH*, Heidenheim, Germany.

Fig. 3.6. Double-sided disc type PM brushless servo motor with the stator ferromagnetic core and built-in brake. Photo courtesy of *Mavilor Motors S.A.*, Barcelona, Spain.

(b) less winding and ferromagnetic core materials for the same torque;
(c) simple stator winding consisting of a single ring-shaped coil (cost effective stator winding, no end connection);
(d) unity winding factor (product of distribution and pitch factor);
(e) the more the poles, the higher the torque density, higher power factor and less the torque ripple;
(f) a three phase motor can be built of three (or multiples of three) identical single-phase units;
(g) a three phase TFM can be fed from a standard three phase inverter for PMBMs using a standard encoder;
(h) motor fed with higher than utility frequency can operate with low speed, and low speed generator can deliver high frequency output current.

Although the stator winding is simple, the motor consists of a large number of pole pairs ($p \geq 24$). There is a double saliency (on the stator and rotor) and each salient pole has a separate *transverse flux* magnetic circuit. Careful attention must be given to the following problems:

(a) to avoid a large number of components, it is necessary to use radial laminations (perpendicular to the magnetic flux paths in some portions of the magnetic circuit), sintered powders or hybrid magnetic circuits (laminations and sintered powders);
(b) the motor external diameter is smaller in the so-called *reversed design*, i.e., with external PM rotor and internal stator (Fig. 3.3a);

76   3 High power density machines

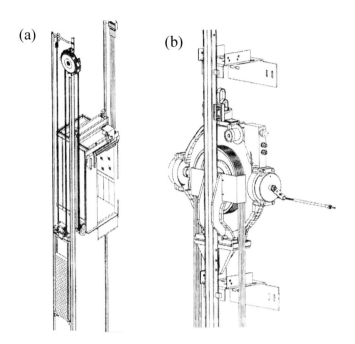

Fig. 3.7. MonoSpace^{TM} elevator: (a) elevator propulsion system; (b) EcoDisc^{TM} motor. Courtesy of *Kone*, Hyvinkaa, Finland.

(c) the TFM uses more PM material than an equivalent standard PM brushless motor;
(d) the power factor decreases as the load increases and special measures must be taken to improve the power factor;
(e) as each stator pole faces the rotor pole and the number of stator and rotor pole pairs is the same, special measures must be taken to minimize the cogging torque.

3.2 Permanent magnet disc type motors

Disc type PMBMs have flat stator and rotor systems in which magnetic flux lines cross the air gap in axial direction (Fig. 3.6). Disc type PMBMs are suitable for electrical vehicles, because they can easily be integrated with wheels or other components of the electromechanical drive system [74]. Low-speed disc type PMBMs are also well suited to gearless elevators, e.g., EcoDisc^{TM}, *Kone*, Hyvinkaa, Finland (Fig. 3.7)[77]. At present time, the highest power density for a 750 kW, 3600 rpm disc type PMBMs with liquid cooling system exceeds 2.2 kW/kg (DRS, Parsippany, NJ, U.S.A.) [74].

3.3 Permanent magnet motors with concentrated non-overlapping coils

The highest *efficiency-to-mass* ratio can be achieved with totally iron free disc type PMBMs described in Section 5.5, provided that PMs are arranged in the so-called Halbach array (Section 2.5.3).

3.3 Permanent magnet motors with concentrated non-overlapping coils

A compact power train of an EV can be designed at minimum costs if a PMBM is designed with concentrated non-overlapping coils. The stator winding coil span is equal to one tooth pitch instead of one pole pitch (Fig. 3.8). Such a winding is similar to the salient pole winding. Owing to very short end connections, the winding losses are reduced, resulting in the increased motor efficiency as compared with a standard PMBM [68, 74].

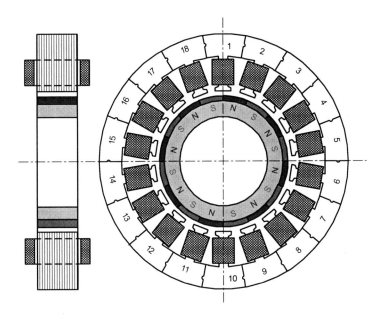

Fig. 3.8. PM brushless machine with concentrated stator coils (stator winding with one slot coil pitch). The stator core is divided into one tooth pitch segments.

The number of armature coils is N_c and the number of poles is $2p$ in a three phase machine must meet the condition $N_c/GCD(N_c, 2p) = 3k$, where GCD is the greatest common divisor of N_c and $2p$ and k is an integer [200]. For example, the following number of coils and poles can be designed: $N_c = 9$, $2p = 6$; $N_c = 12$, $2p = 8$, $N_c = 18$, $2p = 12$, etc. The distribution factor of a single layer non-overlapping winding is always unity, while the pitch factor is affected by the circumferential thickness of the coil [74].

Fig. 3.9. Two-pole, 65 MW, 3600 rpm synchronous motor for LNG plant compressor [102]. Photo courtesy of *Siemens AG*, Erlangen, Germany.

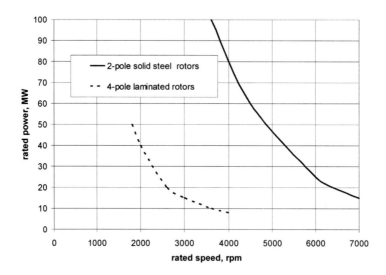

Fig. 3.10. Theoretical limit curves for rating two-pole and four-pole synchronous motors for compressor drives according to *Siemens AG* [102].

3.4 Motors for refrigeration compressors

Refrigeration turbo compressors in modern liquefied natural gas (LNG) plants are traditionally driven by industrial heavy duty gas turbines. With an ongoing industry trend towards larger power train sizes, higher energy efficiency and lower greenhouse gas emissions, the use of very large electric motors to drive the compressors becomes of increasing interest [102]. Gas turbines have the following inherent limitations as compared to equivalent electric motors [102]:

- high thermal and mechanical stresses, which result in lifetime reductions and frequent service requirements of certain components;
- very tight clearances and tolerances between stationary and rotating parts and consequently complexity and sensitivity of turbines;
- relatively poor efficiency and high greenhouse gas emissions;
- no self-starting and acceleration of loaded compressor capability;
- reduced power output at high ambient temperatures;
- poor efficiency at partial load;
- available only as type-tested standardized products with given output ratings and limited speed range;
- limited number of vendors.

Variable speed drives (VSD) with large-power electric motors of equivalent rating do not have the above drawbacks. *Siemens* has built and load-tested refrigeration compressor VSDs rated at 32 to 80 MW and 3600 rpm for LNG plants since 2003 [102]. Fig. 3.9 shows a two-pole cylindrical rotor synchronous motor with brushless exciters for propulsion of a refrigeration compressor. Totally enclosed horizontal water-air cooled (TEWAC) motors are typically used for compressor applications. In some cases, if no cooling water is available on site, totally enclosed air-air cooled motors can be used [102]. Fig. 3.10 shows theoretical limit curves for rating two-pole compressor drive synchronous motors [102].

Installation and operating costs of electric motor VSDs are significantly lower than those of gas turbines. Electric motor drives require very little maintenance and can operate with no power reduction at elevated temperatures.

3.5 Induction motors with cryogenic cooling system

Over the years, various methods of transferring LNG from ship to tank storage or transfer directly from ship into a regasification or send-out system have been studied [163]. Application of a *submerged electric motor pumps* (SEMP) for these types of services is technically and economically justified. For LNG pumps, submerged IMs with cryogenic cooling system rated up to 2.3 MW are used (Fig. 3.11). The temperature of LNG used as a cooling medium is $112K = -161^0C$.

The starting current required is approximately 6.5 times the full load current [163]. It is difficult to reduce this value because of the amount of torque required for starting a cryogenic motor. The rotor, stator core and windings are immersed in LNG. Therefore, the motor is cooled very effectively. Its characteristic feature is large current density at rated load [104]. Active parts (ferromagnetic core and windings) have mass about two times smaller than an

80   3 High power density machines

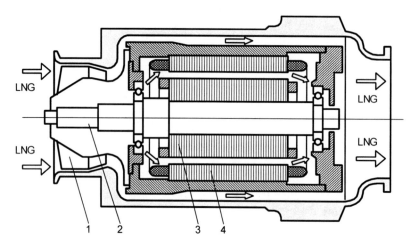

Fig. 3.11. Submerged cage IM with pump. 1 — pump, 2 — shaft, 3 — rotor, 4 — stator [104].

equivalent classical motor of the same rating. To minimize the starting current, soft starters, autotransformers and solid state VFDs can be used with SEMPs. See also Figs 8.22, 8.23 and 8.24.

4
High speed machines

The speed of a.c. machines increases with increase in the input frequency. High frequency of the armature current reduces the dimensions of electrical machines, as the electromagnetic torque is proportional to the electromagnetic power and number of pole pairs and inveresely proportioal to the frequency. High speed gearless electrical machines find many applications as spindle motors, pump motors, large chiller motors, gas compressor motors, microturbine generators and aircraft generators. Elimination of gear trains improves the efficiency of the system, reduces the dimensions and noise, and simplifies the construction and maintenance. Cage induction, wound synchronous and surface type PM synchronous machines with retaining sleeve are the most economical candidates for high speed applications.

At present, the maximum power of high speed synchronous generators does not exceed 500 kW. Several airborne power missions are now evolving that will require lightweight multi megawatt electrical power systems, e.g., directed energy weapon (DEW) and airborne radar [197]. New high power airborne and mobile military systems will require 1 to 6 MW of electrical power generated at speeds 15 krpm. As potential candidates HTS rotor synchronous generators or all cryogenic generators (synchronous or homopolar) have been considered.

4.1 Requirements

Fig. 4.1 shows the construction of a high speed electric machine with magnetic bearings. There are two radial magnetic bearings and one axial magnetic bearing. Basic design requirements for high speed machines include, but are not limited to:

- compact design and high power density;
- minimum number of components;
- ability of the PM rotor to withstand high temperature;
- minimum *cost–to–output power* ratio and *cost–to–efficiency ratio*;

82   4 High speed machines

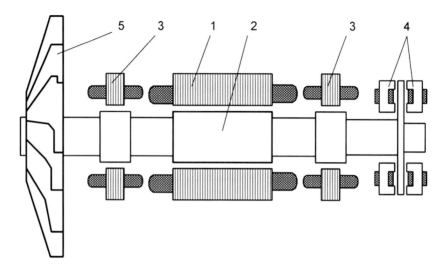

Fig. 4.1. Longitudinal section of a high speed electric machine with magnetic bearings: 1 — stator, 2 — rotor, 3 — radial magnetic bearing, 4 — axial magnetic bearing, 5 — turbine rotor or impeller.

- high reliability (the failure rate < 5% within 80 000 h);
- high efficiency over the whole range of variable speed;
- low total harmonics distortion (THD).

4.2 Microturbines

A *microturbine* (Figs 4.2 and 4.3) is a small, single-shaft gas turbine the rotor of which is integrated with high speed electric generator (up to 120 000 rpm), typically rated from 30 to 200 kW of the output power. In large electric power plants, the turbines and generators are on separate shafts, and are connected by step down gears that slow down the high-speed rotation and increase the torque to turn much larger electric generators.

The stator laminations are about 0.2-mm thick for frequencies below 400 Hz and about 0.1-mm thick for frequencies above 700 Hz. Thin silicon steel laminations (Section 2.1) or sometimes iron-cobalt laminations (Section 2.2) are used for stator and rotor stacks.

The rotor PMs are protected against centrifugal forces with the aid of *retaining sleeves* (cans). The non-magnetic retaining sleeve can be made of non-magnetic metals, e.g., titanium alloys, stainless steels, Inconel 718 (NiCoCr based alloy) or carbon-graphite composites. For metal retaining sleeves the maximum operating temperature is 290^0C and maximum linear surface speed is 250 m/s. For carbon-graphite fiber wound sleeves the maximum operating temperature is 180^0C and maximum linear surface speed is 320 m/s. A

Fig. 4.2. Microturbine set. Photo courtesy of *Capstone*, Chatsworth, CA, U.S.A.

good materials for retaining sleeves have high permissible stresses, low specific density and good thermal conductivity.

Modern generators for distributed generation technologies should meet the following requirements:

- brushless design;
- minimum number of components;
- small volume;
- high power density (output power-to-mass or output power-to-volume ratio);
- high efficiency;
- low cost.

It is also desired that modern brushless generators have more or less fault tolerance capability. However, generating mode with one damaged phase winding and then normal operation after the fault clears is normally impossible.

The first two requirements increase the reliability. Reliability data of older high speed generators are very scattered with mean time between failure (MTBF) values up to approximately 47 000 h as calculated from short-term maintenance record [169].

The higher the speed (frequency) and more efficient the cooling system, the smaller the volume and mass. Increase in speed and application of direct

84   4 High speed machines

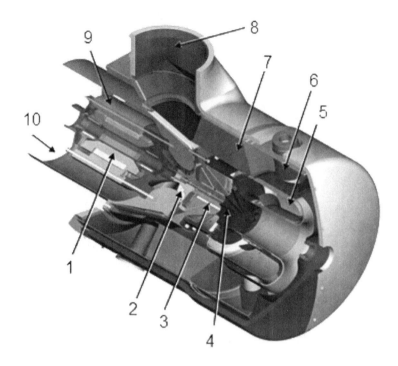

Fig. 4.3. Microturbine with PM brushless generator and air bearings. 1 — generator, 2 — compressor, 3 — air bearings, 4 — turbine, 5 — combustion chamber, 6 — fuel injector, 7 — recuperator, 8 — exhaust outlet, 9 — generator cooling fins, 10 — air intake. Photo courtesy of *Capstone*, Chatsworth, CA, U.S.A.

liquid cooling result in higher power density (output power to mass or output power to volume).

High efficiency means the reduction of the input mechanical power through the reduction of power losses. The lower the losses, the lower the temperature rise of a generator.

Microturbine generators are cooled by the following media:

- air;
- refrigerant;
- oil;
- water.

The air enters through the end bell and passes through the windings and sometimes through rotor channels. The air is exhausted through a perforated screen around the periphery of the casing. Refrigerant is directed to cool the stator core outer surface and/or stator core inner surface (air gap).

The liquid coolant, i.e., oil or water is pumped through the stator jacket or through the stator hollow conductors (direct cooling system) and cooled by means of a heat exchanger system. However, hollow conductors and direct liquid cooling seem to be too expensive for generators rated below 200 kW.

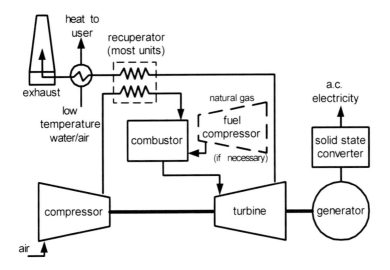

Fig. 4.4. Components of a gas microturbine.

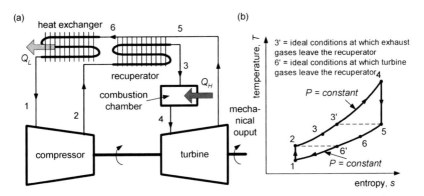

Fig. 4.5. Ideal Brayton cycle modified with recuperation: (a) schematic, (b) temperature – entropy ($T - s$) diagram. Q_H is the high temperature heat transfer rate and Q_L is the low temperature heat transfer rate.

Basic components of microturbines are: turbine compressor, combustor, recuperator, generator and output solid state converter to provide 50 or 60 Hz electrical power (Fig. 4.4).

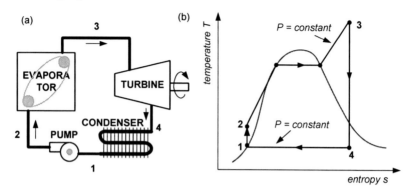

Fig. 4.6. Ideal Rankine cycle: (a) basic components, (b) T–s process.

The most popular microturbines burn natural gas (Fig. 4.4). Outside air is fed into a compressor, which increases the air density and pressure. The compressed air and fuel move into the combustion chamber, where they burn and give off a large amount of heat and high-pressure exhaust gases. The exhaust pushes through a series of turbine rotor blades attached to a long shaft, which drives the shaft at very high speeds. That shaft, in turn, spins the electric generator.

Many of the smaller microturbines are fed by diesel fuel, gasoline or fossil fuels rather than natural gas. In these microturbines there is no need for a compressor, as fuel is injected into the compression chamber.

Some microturbines even include the ability to generate electricity from the heat of the exhaust gases. The heat boils water, and the resulting steam escapes through a second set of turbine blades, spinning a second electric generator. Those systems are much larger and more expensive, but operate more efficiently. Instead of water, an organic substance can also be used, that enters the turbine, where it expands and produces work by rotating the rotor blades.

Despite lower operational temperatures than those of combustion turbines, microturbines produce energy with efficiencies in the 25 to 30% range.

Bryton cycle is a constant–pressure cycle and is generally associated with the gas turbine (Fig. 4.5). The *gas turbine cycle* consists of four internally reversible processes:

(a isentropic compression process;
(b) constant-pressure combustion process;
(c) isentropic-expansion process;
(d) constant-pressure cooling process.

The efficiency of Brayton cycle can be increased with the aid of the so called *recuperation* or *regeneration* (Fig. 4.5a). Recuperation uses the high-temperature exhaust gases from the turbine to heat the gas as it leaves the compressor. The $T - s$ diagram, where T is the temperature and s is the specific *entropy*[1], modified with recuperation is shown in Fig. 4.5b. The recuperation process improves the thermal efficiency of the Brayton cycle because some of the energy that is normally rejected to the surroundings by the turbine exhaust gases is used to preheat the air entering the combustion chamber.

Brayton engine also forms half of the combined cycle system, which combines with a Rankine engine to further increase overall efficiency. The ideal *Rankine cycle* is the model for the steam power plant (Fig. 4.6). It consists of four basic components (Fig. 4.6a):

- pump,
- evaporator (boiler)
- turbine
- condenser

Water is the most common working fluid in the *Rankine cycle*. A disadvantage of using the water-steam mixture is that superheated vapor has to be used, otherwise the moisture content after expansion might be too high, which would erode the turbine blades. Organic substances, that can be used below a temperature of 400°C do not need to be overheated. For many organic compounds superheating is not necessary, resulting in a higher efficiency of the cycle. This is called an *organic Rankine cycle* (ORC).

ORC can make use of low temperature waste heat to generate electricity. At these low temperatures a vapor cycle would be inefficient, due to enormous volumes of low pressure steam, causing very voluminous and costly plants. ORCs can be applied for low temperature waste heat recovery (industry), efficiency improvement in power stations [196], and recovery of geothermal and solar heat. Small scale ORCs have been used commercially or as pilot plant in the last two decades.

Several organic compounds have been used in ORCs, e.g., chloroflourocarbon (CFC), freon, iso-pentane or ammonia to match the temperature of the available waste heat. For example, the R245fa refrigerant is a nonflammable and provides excellent temperature to pressure match.

Combined heat and power (CHP) or *cogeneration* is an energy conversion process, where electricity and useful heat are produced simultaneously in one process. Cogeneration systems make use of the waste heat from Brayton engines, typically for hot water production or space heating. The CHP process may be based on the use of steam or gas turbines or combustion engines.

---

[1] *Entropy* in a closed thermodynamic system is a quantitative measure of the amount of thermal energy not available to do work. Second law of thermodynamics is also called the entropy law.

88   4 High speed machines

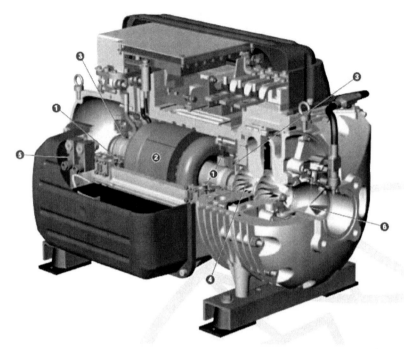

Fig. 4.7. High speed compressor with PM brushless motor: 1 — magnetic bearing, 2 — PM motor, 3 — touchdown bearing (when the compressor is not energized), 4 — shaft and impellers, 5 — compressor cooling, 6 — inlet guide vane assembly. Photo courtesy of *Danfoss Turbocor Compressors*, Dorval, Quebec, Canada.

4.3 Compressors

A high speed compressor with PM brushless motor is shown in Fig. 4.7. The main features are:

- two-stage centrifugal compression;
- high speed PM brushless motor (18 000 to 48 000 rpm);
- impeller integrated with the PM rotor;
- oil-free frictionless PM-assisted magnetic bearings;
- PWM inverter-fed motor;
- power electronics integrated with the onboard intelligent digital electronics;
- sound level less than 70 dBA.

CompAir, Redditch, U.K. manufactures screw-type and reciprocating air compressors in the 1 – 300 kW power range. Its variable speed L45SR, L75SR and L132SR screw air compressors apply SRM drives (produced under license to *SRD*, Harrogate, U.K.). The numbers 45, 75 and 132 indicate the SRM power in kW. The variable speed of a SRM is in the range from 1200 to 5000 rpm.

Fig. 4.8. SRM for a variable speed air compressor. Stator core is not shown. Photo courtesy *CompAir*, Redditch, U.K.

A SRM is shown in Fig. 4.8. These VSD compressors offer the ability to precisely match power consumption with air demand. Field trials show average energy efficiency gain and operational cost savings of over 25% compared to conventional air compressors of the same rating using an a.c. IM and inverter.

4.4 Aircraft generators

The function of the *aircraft electrical system* is to generate, regulate and distribute electrical power throughout the aircraft. Aircraft electrical components operate on many different voltages both a.c. and d.c.. Most systems use 115 V a.c. (400 Hz) and 28 V d.c.. There are several different electric generators on large aircraft (Fig. 4.9) to be able to handle excessive loads, for redundancy, and for emergency situations, which include:

- engine driven a.c. generators;
- auxiliary power units (APU);
- ram air turbines (RAT);
- external power, i.e., ground power unit (GPU).

Each of the engines on an aircraft drives one or more a.c. generators (Fig. 4.10). The power produced by these generators is used in normal flight to

90   4 High speed machines

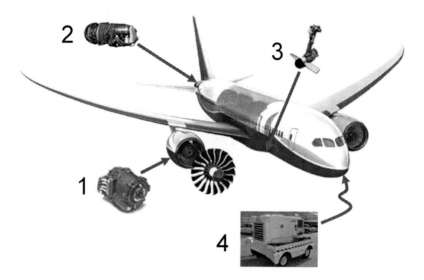

Fig. 4.9. Passenger aircraft generators: 1 — main engine starter/generator, 2 — auxiliary power unit (APU), 3 — emergency ram air turbine (RAT), 4 — ground power unit (GPU).

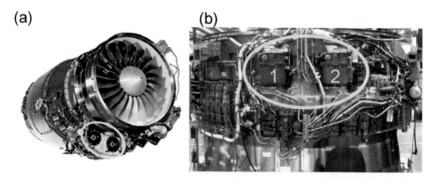

Fig. 4.10. Turbofan engine and engine driven generators (circled): (a) gear trains (generators have been removed); (b) generators (1 and 2). Photo courtesy of United Technologies Corporation, East Hartford, CT, U.S.A.

supply the entire aircraft with power. The power generated by APUs is used while the aircraft is on the ground during maintenance and for engine starting (Figs 4.11 and 4.12). Most aircrafts can use the APU while in flight as a backup power source. RATs are used in the case of a generator or APU failure, as an emergency power source (Fig. 4.13). External power may only be used with the aircraft on the ground. A GPU (portable or stationary unit) provides a.c. power through an external plug on the nose of the aircraft.

4.4 Aircraft generators   91

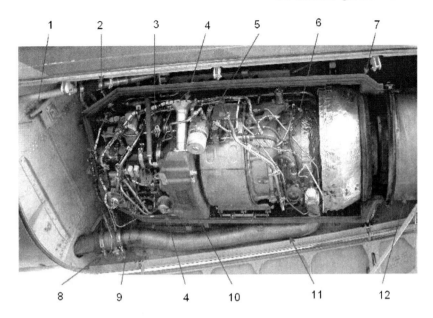

Fig. 4.11. APS 2000 APU of Boeing 737. 1 — light switch, 2 — APU fuel line, 3 — generator, 4 — oil filter, 6 — fuel nozzles, 7 — upper shroud, 8 — bleed air valve, 9 — start motor, 10 — oil tank, 11 — bleed air manifold, 12 — exhaust muffler. Photo courtesy of C. Brady, *The 737 information site* [32].

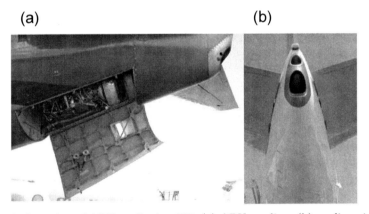

Fig. 4.12. Location of APU on Boeing 737: (a) APU cowling; (b) cooling air inlet above the exhaust. Photo courtesy of C. Brady, *The 737 information site* [32].

Aircraft generators are usually wound rotor synchronous machines with synchronous brushless exciter and PM brushless exciter. The power circuit is shown in Fig. 4.14. PM brushless generators are rather avoided due to difficulties with shutting down the power in failure modes. There are also

Fig. 4.13. Ram air turbine of Airbus A320 located under left wing. Photo courtesy of B. Clayton, www.airlines.net .

attempts of using switched reluctance (SR) generators with no windings or PMs on the rotor. A generator control unit (GCU), or voltage regulator, is used to control generator output. The generator shaft is driven by an aircraft engine with the aid of gears (Fig. 4.10) or directly by low spool engine shaft.

Aircraft generators are typically three-phase synchronous generators with outer stator with distributed-parameter winding and inner rotor with concentrated coil winding (Fig. 4.15). These rules do not apply to special voltage regulated synchronous generators and SR generators. The field excitation current is provided to the rotor with the aid of a brushless exciter.

The stator of synchronous generators has slotted winding located in semiclosed trapezoidal or oval slots. The number of stator slots is typically from 24 to 108, while the number of stator slots per pole per phase is from 4 to 10. Large number of stator slots per pole per phase and double layer chorded windings allow for reducing the contents of higher space harmonics in the air gap magnetic flux density waveforms. At high speeds (high frequency) coils have low number of turns and large number of parallel wires. Very often single turn coils must be designed. The outer surface of the stator core is sometimes serrated to improve the heat transfer from the stator core surface to the stator enclosure or liquid jacket.

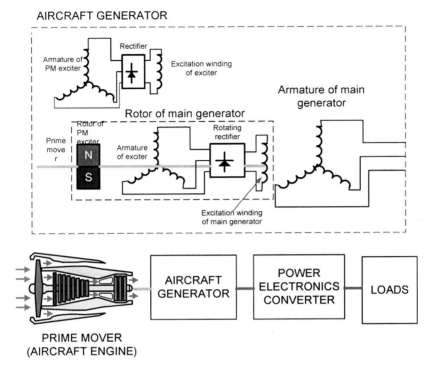

Fig. 4.14. Power circuit of wound rotor synchronous generator for aircrafts.

The number of salient rotor poles is typically from 2 to 12. Pole faces have round semi-closed slots to accommodate the damper. The rotor core is made of the same material as the stator core, i.e., iron-cobalt thin laminations. Rotor coils are protected against centrifugal forces with the aid of metal wedges between poles which also participate in the cooling system of the rotor. Sometimes, in addition to wedges, rotor retaining non-magnetic sleeves are used. With increase of the output power, the rotor cooling problems become very difficult. One of methods is to use aluminum cold plates between the rotor coils and rotor pole core. The rotor inner diameter (shaft diameter) depends amongst other factors on the rotor critical speed. Problems of rotor dynamics are much more serious than in low speed synchronous machines.

The rotor field excitation winding is connected via rotating diode rectifier to a three-phase armature winding of a brushless exciter. The exciter armature system (winding and laminated stack), rectifier and excitation winding of the generator are located on the same shaft. The excitation system of the brushless exciter is stationary, i.e., PMs or d.c. electromagnets are fixed to the stator facing the exciter armature winding. In the case of d.c. electromagnets, the d.c. current can be supplied from an external d.c. source, main armature

Fig. 4.15. Aircraft synchronous generator rated at 90 kW. 1 — stator of main generator with three phase armature winding, 2 — rotor, 3 — stationary field excitation system of exciter, 4 — stator with three phase winding of PM brushless sub-exciter. Photo courtesy of *Hamilton Sundstrand*, Rockford, IL, U.S.A.

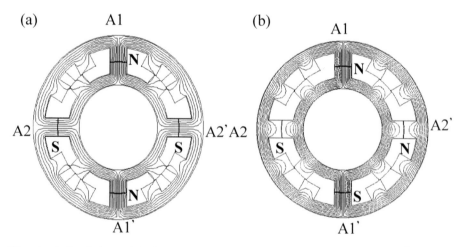

Fig. 4.16. Dual channel high speed SR machine: (a) consequent winding (90^0 magnetic flux path); (b) non-consequent winding (180^0 magnetic flux path) [154].

4.4 Aircraft generators

winding via rectifier, or from a small PM generator (sub-exciter) with stationary armature winding and rotating PMs. Rotating PMs are located on the shaft of main generator.

The frequency of the rotor magnetic flux of a synchronous generator with brushless exciter is speed dependent, i.e., the frequency of the excitation flux decreases as the speed decreases.

Aircraft generators can be driven by the aircraft turbine engine as a prime mover in one of the following way [24, 94, 114, 140, 151, 196],

- engine shaft and generator shaft connected via gear trains;
- engine shaft directly integrated with the generator rotor.

The speed of contemporary aircraft generators is typically from 7200 to 27 000 rpm and output power from 30 to 250 kW.

Both the shaft speed and output frequency of a generator can be constant or variable. Consequently, generators can be divided into the three following groups [24, 94, 153, 196]:

- constant speed constant frequency (CSCF) generators;
- variable speed constant frequency (VSCF) generators;
- variable frequency (VF) generators [2].

A constant output frequency without an a.c. to a.c. utility converter can only be obtained if the generator is driven at a constant speed.

VSCF systems employ an a.c. three-phase generator and solid state converter. The solid state converter consists of (a) a rectifier which converts a variable frequency current into d.c. current, (b) intermediate circuit and (c)inverter which then converts the d.c. current into constant frequency a.c. three-phase current.

In VF systems the output frequency of an a.c. generator is permitted to vary with the rotational speed of the shaft. The variable frequency (VF) is not suitable for all types of a.c. loads. It can be applied directly only to resistive loads, e.g., electric heaters (deicing systems).

Also, generators are turned by a differential assembly and hydraulic pumps to obtain constant speed. The purpose of the constant speed drive (CSD) is to take rotational power from the engine and, no matter the engine speed, turn the generator at a constant speed[3]. This is necessary because the generator output must be constant frequency (400Hz).

An integrated drive generator (IDG) is simply a CSD and generator combined into one unit mounted co-axially or side-by-side.

---

[2] sometimes called 'wild frequency' (WF) generators.
[3] In 1946, adapting technologies developed for machine tools and oil pumps, *Sundstrand Corporation*, Rockford, IL, U.S.A. designed a hydraulically regulated transmission for the *Boeing* B-36 bomber. This CSD converts variable engine speed into constant speed to run an a.c. generator.

A *dual channel SR generator* is a single SR machine that generates the power for the two independent power channels. Each channel has its own power electronics, power EMI filter, and controller, which operate independently and drive separate and independent loads. Fig. 4.16 shows a 12-pole stator and 8-pole rotor SR machine, in which both channels feed two independent solid state converters and receive rotor position information from a single rotor position sensor [154].

Aircraft generators use forced air or oil cooling systems. The most effective is the so called *spray oil cooling* where end connections of stator windings are oil-sprayed. The current density of spray-oil cooled windings can exceed 28 A/mm^2. Pressurized oil can also be pumped though the channels between round conductors in slots.

4.5 High speed multimegawatt generators

4.5.1 Directed energy weapons

Directed energy weapons (DEW) take the form of lasers, high-powered microwaves, and particle beams. They can be adopted for ground, air, sea, and space warfare.

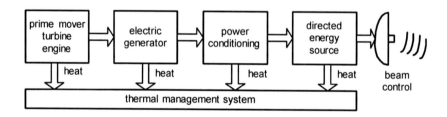

Fig. 4.17. System block diagram for a generic electrically powered airborne DEW system.

Lasers produce either continuous beams or short, intense pulses of light in every spectrum from infrared to ultraviolet. The power output necessary for a weapons-grade high energy laser (HEL) ranges from 10 kW to 1 MW. When a laser beam strikes a target, the energy from the photons in the beam heats the target to the point of combustion or melting. Since the laser beam travels at the speed of light, HELs can particularly be used against moving targets such as rockets, missiles, and artillery projectiles. X-ray lasers may be possible in the not too distant future.

High-power microwave (HPM) weapons produce either beams or short bursts of high-frequency radio energy in the megawatt range. For comparison, a typical microwave oven generates less than 1.5 kW of power. When the microwave energy encounters unshielded current conducting bodies, semiconductors or electronic components, it induces a.c. current in them. The high frequency electric current causes the equipment to malfunction without injuring the personnel. If the energy is high enough, the microwaves can permanently 'burn out' the equipment. The depth of penetration of millimeter-length electromagnetic wave into human skin is very small and does not damage the tissue. Only a burning pain is produced which forces the affected person to escape. Current HPM research focuses on pulsed power devices, which create intense, ultrashort bursts of electrical energy.

A *particle beam* (PB) weapon is a type of DEW which directs an ultra high energy beam of atoms or electrons in a particular direction by a means of particle projectiles with mass. The target is damaged by hitting it, and thus disrupting its atomic and molecular structure. If the target is electric current conductive, a resistive heating occurs and an electron beam weapon can damage or melt its target. Electric circuits and electronic devices targeted by electron PB weapon are disrupted, while human beings and animals caught by the electric discharge of an electron beam weapon are likely to be electrocuted.

There are two technically difficult challenges:

- the high voltage continuous electric power required for DEW systems must be in the range of megawatts;
- a large amount of heat rejected from DEW system during operation must be managed.

The thermal management challenge becomes difficult when the large heat flux is coupled with a small airframe. The electrical power and thermal management subsystem of a conceptual generic airborne electrical DEW system is shown in Fig. 4.17. So far, the electrical power and thermal management systems for airborne DEWs are in early development.

Classical synchronous generators in the range of megawatts would be too heavy for airborne applications. Synchronous generators with HTS rotor excitation windings are investigated as a possible solution. Large power, high speed HTS generators , if available, would be significantly lighter and more compact than conventional copper wire-wound or PM rotor generators.

4.5.2 Airborne radar

Airborne radar systems can be carried by both military and commercial aircrafts and are used for:

- targeting of hostile aircraft for air-to-air combat;
- detection and tracking of moving ground targets;
- targeting of ground targets for bombing missions;
- accurate terrain measurements for assisting in low-altitude flights;
- assisting in weather assessment and navigation;

- mapping and monitoring the Earth's surface for environmental and topological study.

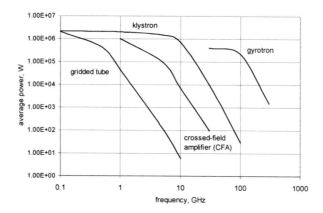

Fig. 4.18. Average output power versus frequency of state-of-the art travelling wave tubes (TWTs).

Radars generally operate in the C or X bands, i.e., around 6 GHz or around 10 GHz, respectively. Airborne radar includes three major categories:

- air-target surveillance and cueing radars mounted in rotodomes;
- nose-mounted fighter radars;
- side-looking radars for ground reconnaissance and surveillance.

The latter is the smallest sector of the airborne radar market and is dominated by synthetic aperture radar (SAR) and ground moving target indicator (GMTI) sensors. SAR, an active all-weather sensor, primarily is used for two-dimensional ground mapping. Radar images of an area help detect fixed targets. GMTI radar picks up moving targets or vehicles. A commercial version of SAR-GMTI, called HiSAR, is an X-band radar that can see from about 100 km away.

The power generation capabilities of *traveling wave tubes* (TWT), i.e., electron tubes used for amplification at microwave frequencies (500 MHz to 300 GHz) range from Ws to MWs (Fig. 4.18). *Klystrons* are the most efficient microwave tubes and are capable of the highest peak and average powers. A klystron is a specialized vacuum tube called a linear-beam tube. The pseudo-Greek word *klystron* comes from the stem form *klys* of a Greek verb referring to the action of waves breaking against a shore, and the end of the word *electron*. Airborne early warning (AEW) systems and weather radars use megawatt klystrons, so that electric generators feeding airborne radars must be rated in MWs range.

4.5.3 Megawatt airborne generator cooling system

Innovative Power Solutions (IPS) has recently announced a new lightweight megawatt-class airborne generator [58, 63]. The size of the generator has been reduced by effective rotor cooling system.

IPS *megawatt airborne generator* is a synchronous generator with salient-pole wound rotor (electromagnetic excitation) and conventional stator with laminated core and winding distributed in slots. A new patented method of cooling the rotor poles and conductors has been implemented [113]. This method uses cold plates disposed between each rotor pole and field coils. A cooling medium (liquid or gas) circulates in the rotor. Each cold plate serves to conduct heat from both the pole core and winding. The cooling medium enters the rotor through the shaft and is distributed between cold plates via manifolds, transfer tubes and plugs. The cooling medium after exiting the rotor (through the shaft) is then conducted to a heat sink or heat exchanger where its temperature is reduced.

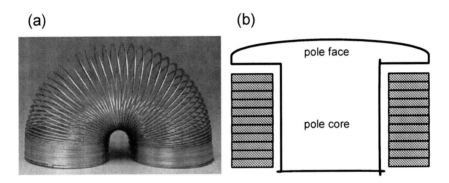

Fig. 4.19. Rotor coil of IPS airborne generator: (a) slinky toy; (b) IPS rotor field excitation coil wound with a flat rectangular conductor.

According to IPS, the lightweight airborne 1 MW generator is 406 mm in diameter, 559 mm long and weighs 210 kg.

To design the rotor field winding, IPS has used flat wires with rectangular cross section in an edge-winding fashion similar to how a slinky toy looks (Fig. 4.19a). The wire is in contact with cooling media along the entire perimeter of the coil. The smaller dimension of the wire is disposed toward the pole core lateral surface and the larger dimension is parallel to the pole face, as shown in Fig. 4.19b. Since a rectangular cross section wire has bigger area of contact between adjacent wires than an equivalent round wire, the heat transfer characteristics for rectangular wires are better.

100   4 High speed machines

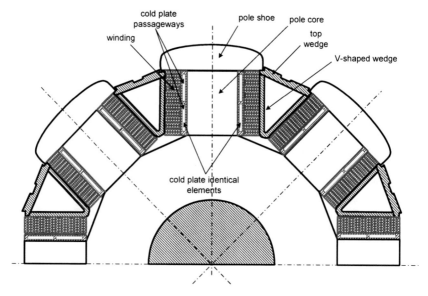

Fig. 4.20. Construction of rotor poles and winding [113].

The rotor may have one or more *cold plates* surrounding each pole core. Fig. 4.20 shows a rotor with a pair of identical cold plates per pole. Each cold plate has passageways for conduction of a cooling medium (Fig. 4.21). Either liquid (oil) or gas cooling medium can be used. The end region of each

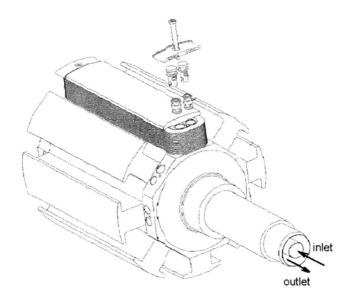

Fig. 4.21. Rotor of 1 MW IPS generator with cooling system [113].

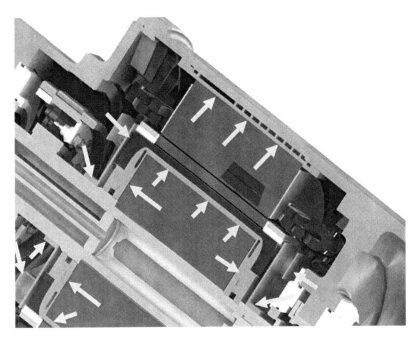

Fig. 4.22. Longitudinal section of IPS' lightweight megawatt generator. Arrows shows cooling locations. Courtesy of *IPS*, Eatontown, NJ, U.S.A.

cold plate matches the bend radius of the field excitation coils. The proposed shape of cold plates does not increase the length and diameter of the rotor.

For fabrication of cold plates high thermal conductivity materials are used, i.e., aluminum, copper or brass. The cold plate preferably includes its own insulating layer, e.g., in the case of aluminum, the insulating material is aluminum oxide with its thickness of 0.125 to 0.25 mm.

To provide the mechanical integrity of the rotor at high speeds and maintain good contact between the winding and cold plates, V-shaped wedges press the winding against cold plate surfaces (Fig. 4.20). Top wedges are used to secure V-shaped wedges in their positions (Fig. 4.20). Cooling locations are shown in Fig. 4.22.

Cold plates can be designed as two-part or single-part cold plates. In the first case both parts are identical. A pair of transfer tubes with plugs at each end of a cold plate provides hydraulic connection with manifolds located at opposite ends of the rotor. This forms a closed system for circulation of cooling medium.

The overall cooling system has been improved by adding radial fans to the rotor and fins to the internal housing. Such a design, although increases windage and ventilation losses, can help to remove heat from the air within the generator and transfer heat to the aluminum housing. Fig. 4.22 shows all cooling locations in IPS megawatt generator. The rotor and stator cooling technique implemented by IPS leads itself to compact generator design; however, the cold plate cooling system is less efficient than spray oil-cooled end windings.

Table 4.1. Selected techniques for enhancing heat dissipation in high speed electric machines.

Cooling system	Current density A/mm2	Advantages	Disadvantages
Fins and heat sinks	5 to 8	Simple method	Increase in weight and size
Water or oil jacket	10 to 15	Effective stator cooling	Increase in diameter and weight
Direct liquid cooling and hollow conductors	up to 30	Very intensive cooling of the stator winding	Increase in weight and size Too expensive for machines rated below 200 kW
Spray oil-cooled end turns of rotor winding	over 28	Very intensive cooling of the rotor winding	Wet rotor; contamination of cooling medium (oil) with time
Liquid cooled wedges [166]	8 to 15 (estimated)	Intensive cooling of rotor winding	Does not effectively cool the rotor poles
Cold plates between poles and rectangular wire rotor winding (IPS)	about 22 (estimated)	Intensive cooling of rotor winding	Requires installation of cold plates in rotor and cooling medium circulation

4.6 Comparison of cooling techniques for high speed electric machines

Table 4.1 shows a comparison of selected *cooling techniques* for high speed electric machines. The current density in the windings depends on the class of insulation, cooling system and duty cycle (continuous, short time or intermittent). The current density values given in Table 4.1 are for $250°$ C maximum operating temperature of windings. The direct cooling system with hollow conductors is the most intensive cooling system (up to 30 A/mm^2). Spray-oil cooling (28 A/mm^2) is almost as intensive as direct cooling. Using cold plates between pole cores and coils the estimated maximum current density should not exceed 22 A/mm^2.

The spray oil-cooled rotor windings allows for maintaining higher current density than cold plates. Spray cooling of the rotor wire together with intensive cooling of the stator winding will theoretically lead to smaller size and weight than application of cold plates.

4.7 Induction machines with cage rotors

It is recommended at high speed to insert cage bars into totally closed rotor slots. Since closed slots tremendously increase the leakage inductance of the rotor winding, the slot closing bridge should be very narrow and saturate when the motor is partially loaded. Instead of closed slots, a narrow slot opening about 0.6 mm can provide a similar effect with moderate rotor winding leakage inductance.

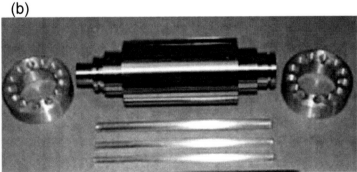

Fig. 4.23. Cage rotor of high speed induction machines: (a) 45 kW, 92 krpm, induction generator; (b) rotor parts for 83.5 kW, 100 krpm induction motor. Photo courtesy of *SatCon*, MA, U.S.A.

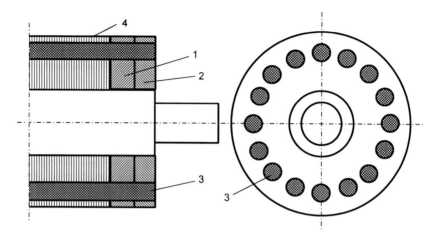

Fig. 4.24. High speed rotor with copper bars, double end rings and laminated stack: 1 — copper or brass end ring, 2 — steel end ring, 3 — copper or brass bar, 4 — laminated rotor stack [56].

It is more difficult to design the rotor end rings than rotor bars. Below are examples of construction of end rings proposed by some manufacturers and researchers.

SatCon, a Massachusetts based company, U.S.A. has extensive experience in the development of high-speed motor and generator systems for a variety of applications [165].

Fig. 4.23a shows a 45kW, 92 krpm, high-speed induction machine developed for the U.S. Army's Combat Hybrid Power Systems (CHPS) program [165]. This machine has been designed for a direct drive generator of a diesel turbocharger for a military ground power application. The linear surface speed of the rotor is 240 m/s. The motor environment was 200°C with 50°C cooling air available to the rotor. A helical stator jacket provides liquid cooling to the stator. High temperature materials have ben required to meet the environmental conditions. The prototype has demonstrated a 97% efficiency. This generator has been equipped with a controlled rectifier to interface with a high voltage bus as part of a highly integrated electrical distribution system for the military vehicle.

Fig. 4.23b shows the components for an 83.5 kW, 100-krpm induction machine for an industrial air compressor [165]. The rotor has closed slots, copper bars and end rings and is integral to the two-stage centrifugal compressor shaft. It is supported on air and magnetic bearings. Similar integrated starter generator (ISG) induction machines have been developed for gas turbine engine applications from 50 000 to 110 000 rpm.

Swiss company *Elektrischemachinen und Antrieb* called shortly *EundA* or *E+A*, Wintertur, Switzerland [56] manufactures laminated rotors with cage windings and composite rotor end rings for high speed induction motors (Fig. 4.24). The outer ring is made of steel, i.e., a material with high radial stress, while the inner ring is made of a high conductivity material, usually copper. The maximum rotor diameter at 60 000 rpm is 65 mm. The maximum linear surface speed is 200 m/s.

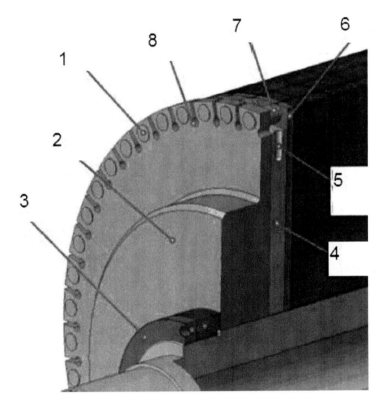

Fig. 4.25. Construction of cage winding of a high speed motor according to US Patent Publication No 2006/0273683A1. 1 — rotor bar, 2 — non-uniform end ring thickness, 3 — clamping nuts, 4 — spacer plate, 5 — balance weight hole, 6 — end laminations, 7 — end ring boss, 8 — keyhole stress relief cut. Courtesy of University of Texas at Austin [36, 37].

Center for Electromechanics at the University of Texas at Austin, TX, U.S.A. has proposed a novel end ring design, which meets all mechanical requirements of high speed, high temperature, and power density without compromising electrical performance [36].

Fig. 4.26. Completed BeCu end ring with integrated joint boss. Courtesy of University of Texas at Austin [36].

Fig. 4.27. Completed end ring to bar solder joints (before trimming extensions). Courtesy of University of Texas at Austin [36].

In a conventional IM rotor design, the end ring is an annular ring supported by the rotor bars. At high operating speeds and temperatures, the centrifugal and thermal growth of the non-self supported end ring would result in high stresses at the bars, laminations, and bar–end ring joints. This configuration also poses a risk of significant mass imbalance due to radial displacement of the unsupported end ring. The 290 m/s linear surface speed of this specific application precludes the use of the low-speed conventional fabricated end ring design [36].

4.7 Induction machines with cage rotors

Fig. 4.28. Completed 2 MW, 15,000 rpm induction motor rotor. Courtesy of University of Texas at Austin [36].

Table 4.2. End ring interference and stress for candidate materials [36]

Quantity	Al	Cu	BeCu
Required radial interference, Pa	58.6	86.2	74.6
Contact pressure at rest, Pa	-144×10^6	-308×10^6	-294×10^6
Contact pressure at operation, Pa	3.65×10^6	1.42×10^6	1.08×10^6
Ring ID hoop stress at rest, Pa	172×10^6	363×10^6	348×10^6
Ring ID hoops stress at operation, Pa	210×10^6	653×10^6	656×10^6
Ring OD radial growth at operation, Pa	157.9	166.9	155.1
Typical yield strength at temperature, Pa	138×10^6	276×10^6	827×10^6

A combination of advanced end ring design features have been developed to alleviate the strength limitations of the end ring–to–bar joint area in the cage rotor assembly for high-speed application, as shown in Fig. 4.25 [37].

Table 4.3. Physical properties of beryllium copper from *NGK Berylco* [139]

Berylco® product	Thermal conductivity W/(m K) at 20°C	Coefficient of linear thermal expansion at 20 to 200°C (length/length)/°C	Modulus of elasticity Pa	Hardness Rockwell (B or C scale)	Tensile strength Pa	Elongation %
Plus	145	18.0×10^{-6}	1.324×10^6	B95 – 102	792.9×10^6	3
Supra	75	17.5×10^{-6}	0.127×10^6	C25 – 32	1172.1×10^6	15
Ultra	60	17.5×10^{-6}	0.127×10^6	C36 – 42	1254.8×10^6	7

In the new design, the end ring is piloted directly to the shaft through an interference fit for rigid support of the end ring to ensure that forces associated with imbalance are not transmitted to the rotor bars (Figs 4.26, 4.27, 4.28). However, at these surface speeds, a uniform cross section end ring is not feasible due to separation of the ring from the shaft resulting from the high centrifugal loads. The end ring was therefore designed with a heavier inner diameter section and a thin web extending to the bar radius. This design maintains compressive interface pressure between the shaft and the end ring throughout the speed and temperature ranges of the machine (0 to 15 000 rpm, −18 to 180°C), with a minimal interference fit that results in manageable stresses. The thick section end ring with direct connection to the shaft serves a secondary purpose of providing bolster support to the laminated stack to prevent conical buckling of the highly interference fitted core necessary for high speed use [36, 37].

Selecting a material for the end ring that balances the electrical and mechanical material requirements was a challenge in this application. Conventional end ring construction (die-cast aluminum and fabricated ETP copper materials) were considered (Table 4.2), but found to be insufficient in strength for this application. Promising recent developments in the use of die-cast copper alloy rotors for high efficiency were reviewed, but still lack the mechanical strength afforded by fabrication with heat treated materials [36].

Beryllium copper (Table 4.3) was selected for adequate strength to withstand the heavy interference fit required to maintain radial contact at the shaft interface during operation at the design speed. Specifically, the selected BeCu C17510 TH04 material provided the best balance between electrical and mechanical requirements. At this heat treat condition, the material has electric conductivity up to 36×10^6 S/m with 668.8 MPa yield and 703.3 MPa ultimate strength [36].

Curtiss-Wright Electromechanical Corporation, PA, U.S.A can manufacture variable speed IMs up to 10 MW and 12 000 rpm [23]. The cage winding surrounds the rotor core. All of the rotor bars are shorted together at each end of the rotor core by full circular conducting end ring. The material of end ring depends on the operation speed. Typically, higher strength copper alloys are used, provided the strength capability can be maintained through the metal joining process. The bar–to–end ring joints are normally accomplished by brazing. In applications where the relatively low-strength copper of the end ring cannot sustain the hoop stress imposed at speed, or the joints cannot accommodate the resulting radial displacements, a high-strength retaining ring is added to provide the necessary support and rigidity. The retaining ring is typically required only in higher speed applications. The retaining ring comprises high-strength alloy steel with good fatigue characteristics. To reduce eddy current losses, the retaining ring should be made of nonmagnetic material. The design of end rings is shown in Fig. 4.29 [23].

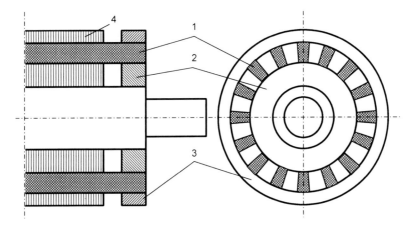

Fig. 4.29. High speed cage rotor winding proposed by *Curtiss–Wright Electromechanical Corporation* , Mount Pleasant, PA , U.S.A. 1 — rotor bar, 2 — end ring, 3 — retaining ring, 4 — laminated or solid steel rotor core [23].

4.8 Induction machines with solid rotors

Research in the area of IMs with solid ferromagnetic rotor were probably initiated in the 1920s by Russian scientists Shenfier [174] and Bruk [34]. In the further years of the 20th Century many researchers and engineers worldwide contributed to the theory and technology of these machines. Major contributions are listed in [50] where detailed analysis of electromagnetic field in these type of IMs has also been presented.

Concepts of solid rotor IMs have been developed in connection with a search for removing drawbacks of cage IMs in order to achieve:

- simplification and reduction of costs of manufacture of the rotor;
- improvement of rotor mechanical integrity at high speed;
- improvement of reliability;
- longer lifetime than wound or cage laminated rotors;
- low vibration and acoustic noise level (in the case of slotless rotor)
- reduction of the inrush starting current of IMs;
- possibility to obtain linear torque-speed characteristic of motors from no load to unity slip due to high solid rotor impedance.

In comparison with cage rotor IMs of the same dimensions, solid rotor IMs have lower output power, lower power factor, lower efficiency, higher no-load slip and higher mechanical time constant. Worse performance characteristics are due to high rotor impedance, higher harmonic eddy currents in solid ferromagnetic rotor body, higher reluctance of solid steel than laminated steel and greater rotor losses due to higher harmonics of the magnetic field than in other types of IMs. There are wide possibilities of reduction of the rotor impedance that improves the performance characteristics through:

110   4 High speed machines

- selecting the rotor solid material with small *relative magnetic permeability –to–electric conductivity* ratio and adequate mechanical integrity;
- using a layered (sandwiched) rotor with both high magnetic permeability and high conductivity materials;
- using a solid rotor with additional cage winding.

Sensible application of the above recommendations that leads to optimization of the design is only possible on the basis of the detailed analysis of the electromagnetic field distribution in the machine. This is why the development of solid rotor machines depends on the advancements in the theory of electromagnetic field in ferromagnetic and non-homogenous structures consisting of materials with different parameters.

Magnetization curves B–H for selected solid steels are plotted in Fig. 4.30. The electric conductivity of solid mild (low carbon) steels is usually from 4×10^6 to 6×10^6 S/m at 20^0C, i.e., 10 to 14 times less than that of copper.

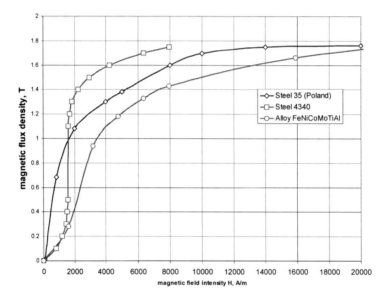

Fig. 4.30. Magnetization curves of various solid steels.

Although the principle of operation of solid rotor IMs is similar to that of other IMs, the analysis of physical effects in solid rotors on the basis of classical electrodynamics of nonlinear bodies is difficult. Problems arise both due to nonlinearity of solid ferromagnetic bodies and complex structures of certain types of these machines. The electromagnetic field in the rotor is strictly three-dimensional (3D) even if the rotating magnetic field excited by the stator system can be assumed as two dimensional (2D). The performance of the machine depends on the intensity and distribution of vectors of this field, in particular, of the vector of current density and the vector of magnetic flux density.

The objective of numerous publications on solid rotor IMs is mostly a formulation of relationships between material parameters, i.e., electric conductivities and magnetic permeabilities, and parameters of the structure, i.e., geometric dimensions and operating performance of a machine under given external conditions on the basis of the electromagnetic field theory.

Recent interest in electric machines with alternating electromagnetic field in solid ferromagnetic rotor parts is motivated by new applications of electrical machines as, for example, motors for high speed direct drive compressors, motors for pumps , motors for drills, high speed generators, electric starters for large turbogenerators, eddy current couplings and brakes, etc. Before the vector control era, there were attempts to use solid rotors covered with thin copper layer for very small diameter rotors of two-phase servo motors, in which it was very difficult to accommodate the cage winding and back iron (yoke). Research is also stimulated by trends in improvements of other type of electrical machines, e.g., machines with rotors made of soft magnetic powder composites (magnetodielectrics and dielectromagnetics), shields of end connections of large turbogenerators, shields for SC machines, retaining sleeve for high speed PM machines and losses in PMs.

Fig. 4.31. Radial turbine, solid rotor coated with copper layer, cooling fan and feed pump [208]. Photo courtsy of the University of Lappeenranta, Finland.

Fig. 4.31 shows a solid rotor of a microturbine developed at the University of Lappeenranta, Finland, for a commercial ORC power plant utilizing the temperature of waste heat [208]. As the relative latent heat of organic fluids is much lower than that of the water, the same or better efficiency as with

Fig. 4.32. Solid rotors with explosive welded copper sleeves for: (a) 300 kW, 63 krpm IM; (b) 3.5 kW, 120 krpm IM. Photo courtesy of *Sundyne Corporation*, Espoo, Finland.

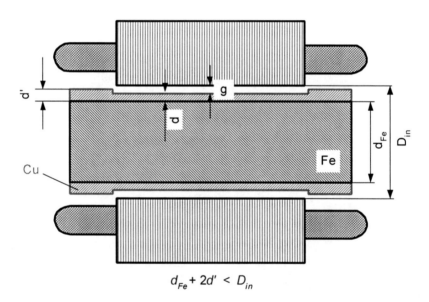

$d_{Fe} + 2d' < D_{in}$

Fig. 4.33. Solid rotor coated with copper layer for high speed induction machines according to U.S. Patent 5473211 [14].

a two-stage steam process can be achieved with a single-stage ORC process, e.g., by using the flue gas heat of a diesel engine. Also, the drop of the specific enthalpy of organic fluids in the turbine is much smaller than that of steam, which makes it possible to make the ORC process efficient at low power.

Solid rotors for high speed induction motors are shown in Fig. 4.32. The rotor construction according to U.S. Patent 5473211 is shown in Fig. 4.33 [14]. The copper layer is thicker behind the stator core than below the stator core, so that the air gap (mechanical clearance) can be minimized.

5
Other types of novel motors

Novel electric motors desribed in this chapter include written pole motors, piezoelectric (ultrasonic) motors, bearingless motors and integrated (smart) motor drives. All these motors are now commercial motors. Electrostatic micromotors have not been considered. This type of motor and actuator has found very limited applications in terms of performance, cost of fabrication and integration with other components of the system (Section 1.5).

5.1 Written pole motor

The stator or a *written pole machine* has a three phase or single phase winding distributed in slots. In addition, there is a concentrated-parameter winding consisting of a single coil placed around an exciter pole (Fig. 5.1). The rotor is furnished with a high resistance cage winding to produce asynchronous torque . The rotor active surface is also coated with a magnetic material layer, which can be magnetized by the stator exciter pole during operation. Any desired magnetic pole pattern can be 'written' on the magnetic layer to provide a hysteresis torque [124]. In this way the starting parameters can be set independently of the operating parameters. The frequency of the exciter winding current is equal to the line frequency (50 or 60 Hz). The machine can be fabricated with either inner rotor (Fig. 5.1a) or outer rotor (Fig. 5.1b).

The written pole motor can develop large synchronous torque below synchronous speed (Fig. 5.2). The hysteresis torque and asynchronous torque start the motor before the exciter pole winding is energized. At subsynchronous speed the exciter winding is energized and writes magnetic poles on the rotor surface. The number of written poles on the rotor surface is inversely proportional to the speed. The rotor is pulled into synchronism when the number of rotor poles is the same as the number of poles of the stator rotating field. By chosing a proper ratio of the exciter current to the stator rotating magnetic field, the rotor can be accelerated with maximum synchronous torque. When the rotor reaches the synchronous speed, the exciter current is switched off

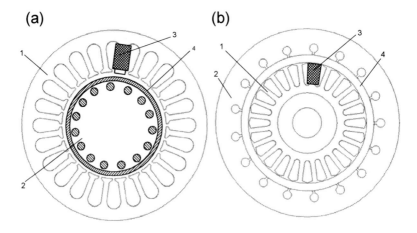

Fig. 5.1. Construction of a written pole machine: (a) with inner rotor; (b) with outer rotor. 1 — stator, 2 — rotor, 3 — stator exciter pole, 4 — rotor ferromagnetic layer.

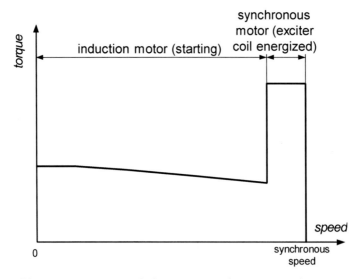

Fig. 5.2. Torque–speed characteristic of a written pole motor.

and the motor operates as a PM synchronous motor (Fig. 5.2). The rotor cage winding designed for high slip (about 20%) reduces the starting current to less than one third of that of a conventional induction motor [124]. At low starting current a written pole motor can accelerate to synchronous speed even with very large moment of inertia.

5.1 Written pole motor

The written pole motor shows the following advantages:

- high efficiency;
- very low start-up current;
- ability to start and synchronize loads with very large inertia;
- ability to operate with about 20% variation in input voltage;
- riding through power outages of up to 15 s;
- restarting after power outages longer than 15 s.

Written pole motors are now available in sizes from 7.5 to 75 kW and are used in remote rural areas with single phase reticulation systems. Table 5.1 contains specifications of single phase written pole motors manufactured by *Precise Power Corporation*, Palmetto, FL, U.S.A. [213]. These motors have low starting current across the line, unity power factor, high power factor during start, smooth (constant torque) start, instant restart after momentary power interruptions and can start with high inertia loads. Single phase written pole motors are ideal for use in rural areas with single-phase reticulation systems.

Table 5.1. Specifications of single phase, high efficiency, internal rotor, TEFC design written pole motors manufactured by *Precise Power Corporation*, Palmetto, FL, U.S.A.

Rated power, kW	22	30	37	45	55	75
Rated power, hp	30	40	50	60	75	100
Frequency, Hz	60					
Speed, rpm	1800					
Rated voltage, V	230	240	460	460	460	460
Full load current, A	104	142	85	103	129	170
Locked rotor current, A	197	282	160	196	225	280
Full load torque, Nm	119	159	201	237	296	410
NEMA iron frame size	365T	445T	445T	445T	445T	449T
Mass, kg	431	816	862	907	998	1226
Ambient starting temperature range	4°C to 40°C					
Efficiency, %	93	93.5	94	94.5	95	95.5

Certain rural applications require high output power motors capable of operating on single-phase power supplies, e.g., oil fields, gas fields, sewage lift stations, crop irrigation, grain drying and handling, lake aeration, roller mills. Written pole motors provide a feasible solution with both economic and environmental benefits.

Written pole motors are used to provide power quality and ride through from utility power sags, surges and interrupts for remotely located weather radars and various aviation applications.

A written pole motor can be combined with a three phase written pole generator. Such a *written pole motor–generator set* has a shared, external, high-inertia rotor and can deliver the highest quality power. Higher harmonics cannot be transferred electromagnetically from the utility grid (motor side) to generator side. The motor–generator sets provide solutions for many three-phase power quality problems, e.g., data reduction, test analysis, laboratories. In industrial manufacturing written pole motor-generator sets are used for process automation computers, distributed information technology (IT) networks, security systems, programmable logic controller (PLCs), telephone systems, data centers. Typical applications in the health care industry are critical care monitoring, intensive care monitoring, nuclear medicine, heart catheter labs, operating rooms and medical records.

5.2 Piezoelectric motors

Piezoelectric motors, also called *ultrasonic motors* operate on the principle of piezoelectric effect which produces mechanical vibrations in the ultrasonic range (Fig. 5.3). The most popular motors are travelling wave piezoelectric motors, invented in Japan in 1982 [204]. The stator has electrodes arranged in two-phase configuration and is fed from a two phase inverter at the frequency above 20 kHz (not audible). The rotor turns in the opposite direction to the travelling wave due to friction between the stator and rotor. Piezoelectric motors produce high electromagnetic torque at low speeds, high holding torque at zero speed, contain only few components and run silently. Typical applications include auto-focus lenses (cameras), timepieces, window blinds, $x - y$ positioning stages, robotics, automobile industry and consumer goods.

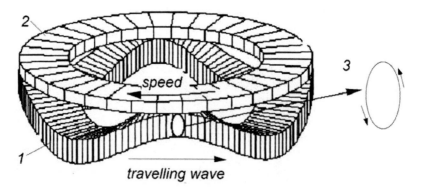

Fig. 5.3. Principle of operation of a travelling wave piezoelectric motor: 1 — stator, 2 — rotor , 3 — orbital of surface particle.

Piezoelectric motors show the following advantages [204]:

- piezoelectric motor can produce a high torque at low speed (Fig. 5.4) with a high efficiency;

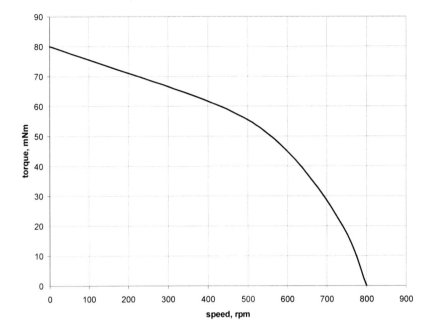

Fig. 5.4. Torque–speed characteristic of a 70 kHz piezoeelectric motor.

- torque density is high;
- inertia of moving piece is low;
- no need for gears since the torque is high;
- no error due to the backlash caused by step down gears;
- the position can be maintained at cutting off the power source due frictional forces between the contact surfaces;
- quiet operation;
- simple construction.

On the other hand, piezoelectric motors require high frequency power sources, strong wear-resistant material and their output power is limited due to the use of vibrational modes.

5.3 Bearingless motors

In a *bearingless motor* the rotor and radial magnetic bearings are integrated. The stator and rotor consist of two parts (Fig. 5.5). Two bearingless rotors are mounted on the same shaft. Each bearingless unit, in addition to tangential electromagnetic forces (electromagnetic torque), produces radial electromagnetic forces. One bearingless unit provides positioning in the x_1 and y_1 directions and the second bearingles units provides positioning the the x_2

120   5 Other types of novel motors

and y_2 directions [41]. Since the electromagnetic torque is produced by two units, the resultant torque is twice the rated torque of each bearingless unit. Each bearingless stator has a three phase suspension winding and three phase torque producing winding (as in standard three-phase a.c. motors). Suspenion windings of two bearingless units are fed from two three phase independent inverters. Suspension currents producing radial forces are commanded by negative feedback controllers and sensors detecting radial position of the shaft. Three phase windings of each stator producing electromagnetic torque are connected in series. Each phase leg contains two series connected windings of the same phase belonging to the first and second stator. Such a winding system is fed from one variable voltage variable frequency (VVVF) inverter. Series connected phase windings and rotors of both units must be correctly aligned. If the stator current corresponds to the q-axis of the rotor in one unit then it also must correspond to the q-axis of the rotor of the other unit [41].

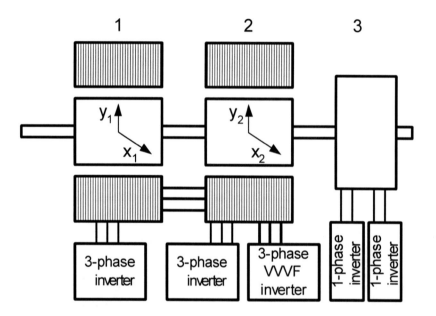

Fig. 5.5. Bearingless motor: 1 — bearingless unit I, 2 — bearingless unit II, 3 — thrust magnetic bearing.

Bearingless motor drives, in comparison with motors with magnetic bearings, show the following advantages [41]:

- compact construction;
- short shaft which results in high critical speeds and stable operation;
- low cost of the system due to reduced number of inverters and connecting wires;
- high power because the electromagnetic torque is produced by two units.

Bearingless induction motors have found so far a few applications as canned pump motors (15 and 30 kW), blood pump motors (Section 6.4) and computer disc drive motors.

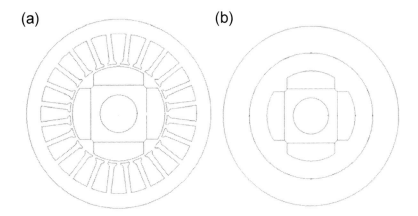

Fig. 5.6. Outlines of magnetic circuits of PMBMs with: (a) slotted stator; (b) slotless stator. Images produced with the aid of SPEED software, University of Glasgow, U.K.

5.4 Slotless motors

In a standard PMBM, conductors are placed in slots cut in the stator ferromagnetic core (Fig. 5.6a). In a *slotless design* (Fig. 5.6b) conductors are distributed on the inner surface of the stator core and held with adhesive. The slotless winding can be a basket type winding inserted into a laminated ring and encapsulated. Slotless design provides:

- smooth rotation without cogging torque [1]
- elimination of losses in PMs and conductive retaining sleeves (if exist) caused by slot space harmonics;
- sinusoidal distribution of magnetic flux density in the air gap due to lack of the stator core teeth;
- lower acoustic noise of electromagnetic origin;
- low stator winding inductance (0.1 to 0.01 of that of slotted PMBM) due to large air gap;
- better motor dynamic response;
- lighter design.

---

[1] Cogging torque is produced as a result of interaction of rotor PMs and stator ferromagnetic teeth. Cogging torque is produced also at zero current state.

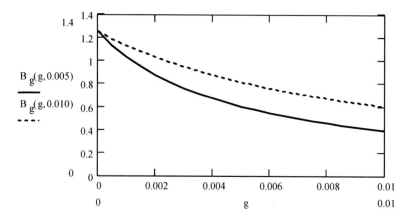

Fig. 5.7. Magnetic flux density in the air gap (T) as a function of air gap (from 0 to 0.01 m) for two radial heights of NdFeB PM: 0.005 m (solid line) and 0.010 m (dash line). The remanent magnetic flux density of PM is 1.25 T and coercivity 920 kA/m.

The main drawback of slotless PMBMs is much larger amount of PM materials necessary to provide the same magnetic flux density in the air gap as that in slotted PMBMs. The air gap magnetic flux density decreases as the air gap increases. The radial height of PMs in slotless motors must be adequate to obtain the required magnetic flux density in large non-magnetic air gap (Fig. 5.7).

Slotless PMBMs are used in:

- medical metering and pumping devices to precisely deliver fluids into delicate areas, e.g., such as eyes;
- medical imaging equipment to provide smoother operation at low speeds;
- surgical handpieces ;
- aircraft flight control systems to provide smoother feedback to pilots;
- hand-held production tools to reduce ergonomic problems by eliminating cogging effect;
- material handling equipment;
- disc and tape drives;
- robotics and robot grippers;
- blowers and fans;
- printers and copiers;
- computer peripherals.

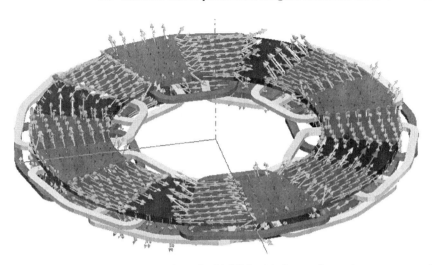

Fig. 5.8. Iron free disc type eight-pole PMBM with inner three-phase stator and twin Hallbach array PM rotor. Author's 3D FEM simulation.

Fig. 5.9. AFPM brushless e-$TORQ^{TM}$ motor with coreless stator windings: (a) general view; (b) motor integrated with wheel of a solar powered car. Photo courtesy of *Bodine Electric Company*, Chicago, IL, U.S.A.

5.5 Coreless stator permanent magnet brushless motors

5.5.1 Disc type coreless motors

Coreless stator PMBMs have their stator windings without any ferromagnetic cores. Disc type motors can totally be iron free motors (Fig. 5.8) provided that PMs are arranged in the so–called Halbach array [78, 79, 80]. Such a motor has neither the armature nor rotor core. The internal stator consists of full-pitch or short-pitch coils wound from insulated wires. Coils are arranged in overlapping layers like petals around the center of a flower and embedded in a plastic of very high mechanical integrity. The stator winding can also consist of concentrated non-overlapping coils (Section 3.3). The twin non-magnetic rotor discs have cavities of the same shape as PMs. Magnets are inserted in these cavities and glued to the rotor discs. Ironless motors have very high efficiency (no core losses), do not produce any torque ripple at zero current state and are lightweight motors. The drawback is larger amount of PM materials and lower stator winding inductance as compared with PMBMs with ferromagnetic cores. Specifications of disc type PMBMs with coreless stators are given in Table 5.2.

Table 5.2. Specifications of $e\text{-}TORQ^{TM}$ disc type PMBMs with coreless stators manufactured by *Bodine Electric Company*, Chicago, IL, U.S.A.

Parameter	178-mm AFPM motor		356-mm AFPM motor	
	Low EMF constant	High EMF constant	Low EMF constant	High EMF constant
Output power, kW	0.7	0.57	1.0	0.26
Number of poles	8	8	16	16
d.c. bus voltage, V	170	300	170	300
Speed, rpm	3000	1500	300	70
Peak speed, rpm	3500	2200	700	400
Torque, Nm	2.26	2.83	31.1	33.9
Peak torque, Nm	22.6	13.67	152.1	84.4
Current, A	5	2	13	5
Peak current, A	50	10.5	64	12.5
Efficiency, %	81	75	84	77
Torque constant, Nm/A	0.4859	1.299	2.38	6.814
EMF constant, V/krpm	50	137	249	713
Winding resistance, Ω	2.2	14.3	1.33	43
Winding inductance, mH	1.4	10.5	3.6	29.4
Viscous friction, Nm/rad/s	9.9×10^{-5}	0.00019	0.00669	0.012
Static friction, Nm	0.00728	0.0378	0.02357	0.1442
Electrical time constant, ms	0.6364	0.734	2.71	0.684
Mechanical time constant, ms	4.538	4.026	4.5	17.78
Moment of inertia, kgm^2	0.00525	0.00525	0.21	0.21
Mass of active parts, kg	6.17	6.17	30.87	30.87
Power density, W/kg	113.5	92.4	32.39	8.42

High efficiency coreless disc-type PMBMs are excellent machines for application where efficiency is the primary demand, e.g., solar powered vehicles (Fig. 5.9), solar powered boats or even *solar powered aircrafts* (Fig. 5.10).

Table 5.3. Specifications of a three-phase, 160-kW, 1950-rpm disc type brushless synchronous generator with ironless stator core built at the *University of Stellenbosch*, Stellenbosch, South Africa [74].

Design data	
Output power P_{out}, kW	160
Speed n, rpm	1950
Number of phases m_1	3 (Wye)
Rated line voltage, V	435 V
Rated phase current, A	215 A
Frequency f, Hz	100
Number of stator modules	1
Number of pole pairs p	20
Number of stator coils (3 phases)	60
Number of turns per phase	51
Wire diameter, mm	0.42
Number of parallel wires a_w	12
Axial height of PM per pole h_M, mm	10.7
Axial thickness of the winding t_w, mm	15.7
Air gap (one side) g, mm	2.75
Air gap magnetic flux density B_{mg} under load, T	0.58
Current density, A/mm^2	7.1
PM outer diameter D_{out}, mm	720
$k_d = D_{in}/D_{out}$ ratio	0.69
Class of insulation	F
Winding temperature rise, °C	56
Cooling system	Self air-cooled
Type of winding	Single layer trapezoidal

Solar powered stratospheric high altitude (17 to 22 km) airborne apparatus are proposed for surveillance, positioning, navigation, remote sensing, and broadband communications. The aircraft power train consists of solar cells placed on the wings, the energy storage system (fuel cell stack), the energy management electronics, lightweight brushless PM motors, and propellers. *Pathfinder*, *Centurion* and *Helios* solar aircrafts have been developed by *Aerovironment Inc.*, Monrovia, CA, U.S.A., as part of the NASA Environmental Research Aircraft and Sensor Technology (ERAST) program at the Dryden Flight Research Center, Edwards, CA, U.S.A. *Helios* has a wing span of 76.8 m, weighs 727 kg, can take 100 kg of payload and is propelled by eight to fourteen 1.5 kW PMBMs. The cruise speed is 30 to 43 km/h and ground speed is 275 km/h. Photovoltaic cells provide up to 22 kW of power at high noon on a clear summer day. The solar cells that cover the entire upper surface of the *Helios* wing span does not only power the electric motors during the day, but also charge the on-board fuel cells that will power the aircraft during night flying. Oxygen and hydrogen gases are accumulated in separate tanks.

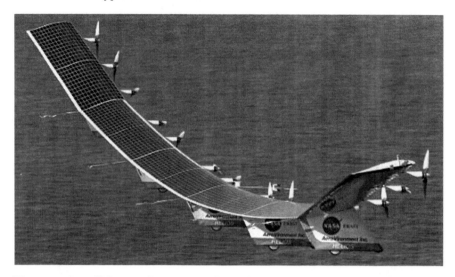

Fig. 5.10. *Helios* solar powered aircraft. NASA photo ED01-0209-3, www.dfrc.nasa.gov/gallery/Photo/Helios/index.html

At night, when the solar cells stop producing electricity, the oxygen and hydrogen gases are fed back into the fuel cell to produce water and electricity for motor operations untill the next morning, when the charging process starts over again. The flight altitude depends on the mission and it is typically from 15 000 to 22 000 m. Maximum flight altitude is 30 500 m. All main parts of *Helios* are made of carbon fibres and styrofoam.

Disc type coreless PM brushless machines can also be used as wind generators (Table 5.3). The number of phases is usually three or sometimes more.

5.5.2 Cylindrical type motors with coreless stator winding

Cylindrical type PMBMs developed and manufactured by *ThinGap Corporation*, Ventura, CA, U.S.A. are moving magnet, slotless, stationary armature winding motors (Fig. 5.11). The PM rotor is designed with ferromagnetic return paths for the magnetic flux. Ring-shaped silicon-steel laminations, shaft and PMs form a rotor [76, 161]. The ring-shaped laminations provide a path for magnetic flux, but do not have any teeth [76, 161]. The stator winding is made from precision-milled sheet copper in a fiberglass and polyimide material composite (Fig. 5.12).

Coreless slotless cylindrical-type PMBMs overcome many ferromagnetic core motor weaknesses. In standard motors, copper occupies only half of the slot winding space, because the slot fill factor is about 0.5. Slot openings modulate the air gap magnetic-flux density waveform and thus, create additional higher space harmonics. On the other hand, the large non-magnetic air gap of a slotless motor requires larger volume of PM material.

5.5 Coreless stator permanent magnet brushless motors

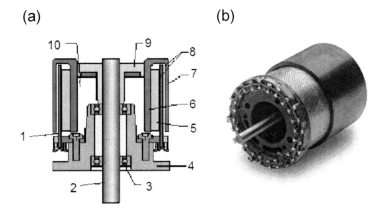

Fig. 5.11. PMBM with coreless stator winding: (a) longitudinal section; (b) partially disassembled motor. 1 — stator winding, 2 — shaft, 3 — bearing, 4 — base, 5 — PM, 6 — inner rotor core, 7 — outer rotor core, 8 — air gap, 9 — hub, 10 — open space. Photo courtesy of *ThinGap*, Ventura, CA, U.S.A.

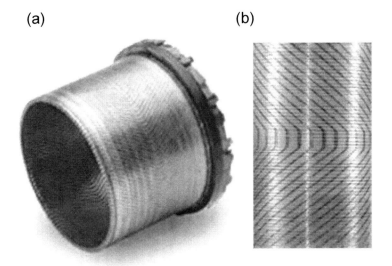

Fig. 5.12. Stator (armature) winding of a coreless stator PM brushless motor: (a) winding in a fiberglass and polyimide material composite; (b) precision-milled sheet copper coils. Photo courtesy of *ThinGap*, Ventura, CA, U.S.A.

Table 5.4 shows specifications of *ThinGap* PMBMs. The TG2320 motor rated at 1025 W, 16.3 krpm shows the highest power density, i.e., 2.19 kW/kg. However, the efficiency of this motor is low, i.e., the maximum efficiency at 80 V d.c. and 1025 W shaft power is only 73%.

Table 5.4. Specifications of coreless PMBMs manufactured by *ThinGap*, Ventura, CA, U.S.A.

	TG2300-ENC		TG2310	TG2320	TG2330	
Continuous rating	12 V	48 V	24 V	80 V d.c.	24	66
•shaft power, W	66	266	182	1025	210	813
•speed, rpm	1230	7200	2734	16 305	2710	8650
•torque, Nm	0.508	0.353	0.636	0.60	0.741	0.897
•current, A	9.2	6.5	12.0	17.65	12.3	15.0
Peak torque, Nm	4.05		3.18	3.0	4.24	
Peak current, A	70		70	86.7	67	
No load speed, rpm	4060		3900	21 800	3700	11 000
Maximum locked rotor torque, Nm	not given		not given	0.32	0.516	
Rotor inertia, kgm^2	1.843×10^{-4}		1.843×10^{-4}	1.631×10^{-4}	3.898×10^{-4}	
Motor mass, kg	1.0		0.547	0.468	0.737	
Torque density, Nm/kg	0.508	0.353	1.162	1.282	1.005	1.217
Power density, kW/kg	0.066	0.266	0.332	2.19	0.285	1.103
Maximum system efficiency, %	not given		not given	73	90	

The slotless and coreless rotor construction and very low saturation of *ThinGap* motors produce an inherently low harmonic distortion in the air gap magnetic flux density and EMF waveforms.

The inductance of *ThinGap* motors is very low. A standard, off-the-shelf amplifier will not provide the best low-error correction. Adding an inductor between the motor and amplifier will not necessary improve the situation where ultra low velocity errors are required. The inductor creates propagation delays that make precision speed control of 0.0001 % or less unmanageable. Any deviation of the back-EMF or current waveforms from a perfect sinusoidal signal causes torque ripple in conventional PMBMs. The *ThinGap* motor with its sinusoidal EMF waveform generates constant torque when using a sinusoidal current. Matching the motor with a sinusoidal–drive amplifier produces the proper coupling of waveforms that results in extremely low torque ripple.

According to *ThinGap*, the precision-milled sheet copper winding technology delivers enhanced performance and offer several advantages, e.g.,

- high power density;
- no hysteresis torque;
- no cogging torque;
- no radial magnetic pull between the rotor and stator;
- smooth quiet operation;
- low inductance coil;
- low magnetic saturation (large air gap);
- low contents of EMF harmonics;

- enhanced heat dissipation due to both sides of the coil exposed to airflow;
- manufacturing precision (automated, repeatable process control).

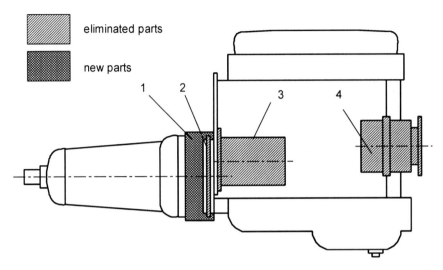

Fig. 5.13. Comparison of classical starter and generator with ISG for motor vehicles. 1 — ISG, 2 — flywheel, 3 — classical starter, 4 — classical generator (alternator).

5.6 Integrated starter generator

The *integrated starter generator* (ISG) replaces the conventional starter, generator and flywheel of the automotive engine, integrates starting and generating in a single electromechanical devices and provides the following auxiliary functions:

- Automatic vehicle start-stop, which switches off the combustion engine at zero load (at traffic lights) and automatically restarts the engine in less then 0.3 s when the gas pedal is pressed.
- Pulse-start acceleration of the combustion engine to the required cranking idle speed and only then the combustion process is initiated. Automatic start-stop system and pulse-start acceleration improves the fuel economy up to 20 % and reduces emissions up to 15 %.
- Boost mode operation, i.e., the ISG operates as electric auxiliary motor to shortly drive or accelerate the vehicle at low speeds.
- Regenerative mode operation, i.e., when the vehicle brakes, the ISG operates as electric generator, converts mechanical energy into electrical energy and helps to recharge the battery.

Fig. 5.14. Induction ISG and solid state converter. Photo courtesy of *Siemens*.

- Active damping of torsional vibration which improves driveability.

The direct drive ISG combines functions of the classical generator (alternator) starter, flywheel , pulley, and belt (Fig. 5.13). Induction ISG and associated power electronics is shown in Fig. 5.14.

5.7 Integrated electromechanical drives

The *integrated!electromechanical drive* contains the electromechanical, electrical and electronic components, i.e., motor, power electronics, position, speed and current sensors, controller, communication integrated circuit (IC) and protection system in one enclosure. Integrated motor drives have the following advantages:

(a) the number of input wires to the motor is reduced;
(b) traditional compatibility problems are solved;
(c) standing voltage wave between the motor and converter increasing the voltage at the motor terminals is reduced;
(d) installation is simple.

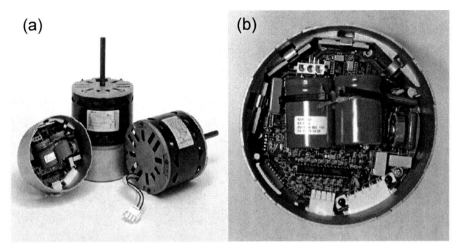

Fig. 5.15. Small integrated induction motor drive: (a) motor; (b) power and control electronics. Photo courtesy of *AO Smith*, Milwaukee, WI, U.S.A.

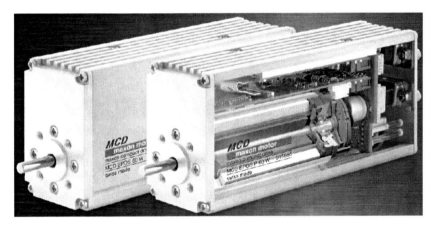

Fig. 5.16. Integrated intelligent PMBM drive. Photo courtesy of *Maxon Precision Motors*, Fall River, MA, U.S.A.

Small integrated induction motor drive is shown in Fig. 5.15. A compact integrated electromechanical drive consisting of PMBM with slotless winding, encoder and controller and offering intelligence is shown in Fig. 5.16.

5.8 Induction motors with copper cage rotor

Until recently, *die cast aluminum cage* rotors have been manufactured because copper pressure die-casting was unproven. Lack of a durable and cost effective

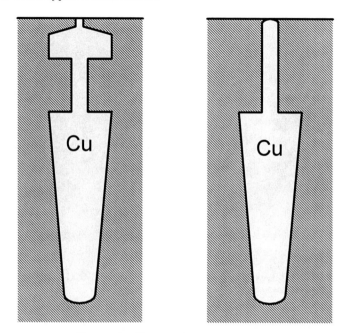

Fig. 5.17. Shapes of copper cage bars.

mold material has been the technical barrier preventing manufacture of the *copper cast cage* rotor. The high melting point of copper (1084.6^{0}C) causes rapid deterioration in dies made of traditional tool steels. Studies conducted by the International Copper Research Association (INCRA) in the 1970s has identified tungsten and molybdenum as good candidate materials for copper casting. They have not found use in the industry largely because of fabrication costs. In the late 1990s, several new materials were developed for high temperature applications. Two promising candidates for parts of the die caster are the nickel-based superalloys and the beryllium-nickel alloys. None of these materials has the low expansion of tungsten or molybdenum, but they do retain exceptional strength at high temperatures. Inconel 617, 625 and 718 alloys operated in the 600 to 650^{0}C temperature range are very promising die materials for die casting of copper motor rotors.

The use of copper in place of aluminum for the cage rotor windings of induction motors reduces the rotor winding losses and improves the motor efficiency. The overall manufacturing cost and motor weight can also be reduced.

The higher conductivity of copper increases the skin effect (decrease in the depth of penetration) and therefore makes deep bar and double cage effects more pronounced. To increase the bar resistance at high frequency (unity slip at starting) the cross section of the rotor bar should be as shown in Fig. 5.17 [192].

5.8 Induction motors with copper cage rotor

Table 5.5. Test data and performance of 1.1, 5.5, 11 and 37 kW IMs. Standard efficiency series aluminum rotor models compared to high efficiency copper rotor designs at 400 V, 50 Hz [98].

Rated power, kW	1.1		5.5		11		37	
Rotor conductor	Al	Cu	Al	Cu	Al	Cu	Al	Cu
Rated current, A	2.68	2.45	11	10.9	21.8	21.9	67.1	67.5
Power factor	0.77	0.79	0.83	0.83	0.83	0.81	0.87	0.85
Speed, rpm	1418	1459.5	1424	1455.7	1437	1460	1468	1485
Rated torque, Nm	7.4	7.21	36.9	36.15	73	71.9	240	237.7
Slip	0.055	0.027	0.051	0.0295	0.042	0.0267	0.021	0.01
Input power, W	1435	1334	6485	6276	12590	12330	40700	39900
Stator winding losses, W	192.6	115.1	427.4	372.4	629	521	1044	975
Core losses, W	63.6	51	140.8	101	227	189	749	520
Stray load losses, W	9.5	6.7	100.3	31.4	163	171	699	200
Rotor losses, W	64.1	31.4	299.2	170.4	483	311	837	451
Windage and friction, W	15.9	25	17.5	36	63	56.5	304	203
Efficiency, %	75.9	82.8	84.8	88.12	87.6	89.9	91.1	93.2
Temperature rise, K	61.1	27.8	80.0	61.3	75	62.1	77.0	70.4

Comparison of IMs with aluminum cage rotors and copper cage rotors is shown in Table 5.5 [98].

Nowadays, motors with copper cage rotors are being produced in sizes ranging from 100 W to almost 100 kW.

6
Electric motors for medical and clinical applications

When electricity was new, people had high hopes that it had curative powers. For example, *electropathy* (electrodes between patient's hands and ailing body part), very popular from 1850 to 1900, promised to cure most diseases and conditions, including mental illness. The 21st Century biomedical engineering community has resurrected magnetic fields to treat depression, e.g., *magnetic seizure therapy* (high frequency, powerful electromagnets) or *transcranial magnetic stimulation* (strong pulse magnetic fields).

At present, many medical devices use small PM electric motors as, for example, patient handling equipment (hospital beds, surgical tables, stretchers, hoists), high-quality pumps, centrifuges, infusion pumps , insulin pumps , hemodialysis machines, compressors , respirators, scanners, dental drills, precision surgical handpieces , surgical robot grippers, prosthesis and implantable devices (ventricular assist devices, pacemakers, defibrillators, nerve stimulators, etc). Also, electric motors are widely used in healthcare equipment, e.g., exercisers, wheelchairs, massage apparatus, therapy equipment, etc.

This chapter focuses on very small modern PM brushless motors for implantable devices (motors for rotary blood pumps), catheters, capsule endoscopes, minimally invasive surgical tools and robots.

6.1 Electric motors and actuators

The most important choice for medical device motor designers is between ferromagnetic core and coreless motors [143]. The elimination of the heavy ferromagnetic core offers such advantages as reduced mass, low electrical time constant, high power efficiency, zero cogging torque and low acoustic noise. Ferromagnetic core-free motors have become the rule in battery-operated or remotely situated devices where rapid cycling or long battery life is important.

Since the reliability of medical products is critical, the motor is considered a precision component rather than a commodity device [143]. Motors and actuators for medical applications must frequently endure hostile environments,

Table 6.1. Miniature PMBMs manufactured by *Faulhaber*, Schonaich, Germany [194].

Specifications	Motor type		
	001B	006B	012B
Diameter of housing, mm	1.9	6.0	6.0
Length of housing, mm	5.5	20.0	20.0
Rated voltage, V	1.0	6.0	12.00
Rated torque, mNm	0.012	0.37	0.37
Stall torque, mNm	0.0095	0.73	0.58
Maximum output power, W	0.13	1.56	1.58
Maximum efficiency, %	26.7	57.0	55.0
No–load speed, rpm	100 000	47 000	36 400
No–load current, A	0.032	0.047	0.016
Resistance line-to-line, Ω	7.2	9.1	59.0
Inductance line-to-line, μH	3.9	26.0	187.0
EMF constant, mV/rpm	0.00792	0.119	0.305
Torque constant mNm/A	0.0756	1.13	2.91
Rotor moment of inertia, gcm^2	0.00007	0.0095	0.0095
Angular acceleration, rad/s^2	1350×10^3	772×10^3	607×10^3
Mechanical time constant, ms	9.0	6.0	6.0
Temperature range, ^{0}C	-30 to $+125$	-20 to $+100$	-20 to $+100$
Mass, g	0.09	2.5	2.5

caustic fluids, radiation, steam, elevated temperatures, vacuum, vibration and mechanical impact.

Specifications of very small PMBMs for surgical devices, motorized catheters and other clinical engineering devices are listed in Table 6.1 and Table 6.2. Small 13-mm diameter, 73.6-mNm, 50-V d.c. PMBM for power surgical and dental instruments is shown in Fig. 6.1. It can withstand in excess of 1000 autoclave[1] cycles. The speed-torque characteristic of a 6-mm, 1.58 W PMBM is shown in Fig. 6.2.

Actuation technologies are compared in Table 6.3 [103, 105]. The shape memory wire is actuated by passing current through it and heating it to about 75^{0}C. The highest power density is for magnetostrictive, piezoelectric and voice coil actuators (VCA).

Improvements in motor capabilities have already helped medical device manufacturers bring some products as portable instruments and surgical tools to market [143]. Computer numerical control production and assembly equipment has enabled motor manufacturers to hold tolerances in the micrometer range, maintain consistency from piece to piece, and customize products in small quantities.

---

[1] Autoclave is an apparatus (as for sterilizing) using superheated high-pressure steamcycles.

Table 6.2. Small BO512-050 PMBMs for power surgical and dental instruments manufactured by *Portescap*^{TM}, *A Danaher Motion Company*, West Chester, PA, U.S.A.[121].

Specifications	Motor type	
	Design A	Design B
Diameter of housing, mm	12.7	
Length of housing, mm	47.0	
Rated d.c. voltage, V	50	50
Peak torque, mNm	73.6	36.5
Stall torque, mNm	7.34	7.34
Maximum continuous current, A	1.13	0.56
Peak current, A	11.3	2.79
No–load speed, rpm	70 100	35 000
Resistance line-to-line, Ω	4.27	4.25
Inductance line-to-line, mH	0.22	0.89
EMF constant, V/krpm	0.68	1.37
Torque constant mNm/A	6.57	13.1
Rotor moment of inertia, 10^{-8} kgm^2	4.94	4.94
Mechanical time constant, ms	7.45	7.51
Electrical time constant, ms	0.05	0.05
Thermal resistance, ^{0}C/W	15.9	15.9
Mass, g	44	44

Table 6.3. Comparison of different actuation technologies [103, 105].

Actuator type	Maximum strain %	Maximum pressure MPa	Maximum efficiency %	Relative full cycle speed	Power density
Voice coil	50	0.10	> 90	Fast	High
Piezoelectric					
• ceramic (PZT)	0.2	110	> 90	Fast	
• single crystal (PZN-PT)	1.7	131	> 90	Fast	High
• polymer (PVDF)	0.1	4.8	n/a	Fast	
Electrostatic devices (integrated force array)	50	0.03	> 90	Fast	Low
Shape memory polymer wire	100	4	< 10	Slow	Medium
Thermal expansion	1	78	< 10	Slow	Medium
Magnetostrictive (Terfenol-D)	0.2	70	60	Fast	Very high

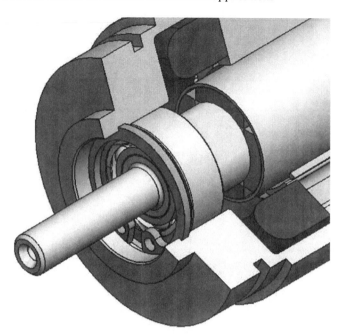

Fig. 6.1. Small 13-mm diameter, 73.6-mNm, 50-V d.c. BO512-050 PMBM (Table 6.2). Photo courtesy of *Portescap^{TM}, A Danaher Motion Company*, West Chester, PA, U.S.A.

6.2 Material requirements

PMs for medical devices should have high energy density, increased oxidation resistance and stable magnetization curves over extended periods of time. Miniaturization of brushless motors is possible due to availability of modern NdFeB PMs with remanence B_r up to 1.45 T, coercivity H_c over 1100 kA/m and $(BH)_{max}$ product about 400 kJ/m^3.

Compared with popular sintered bronze bearings and expensive stainless-steel ball bearings, sintered ceramic bearings provide up to 50% more load-bearing capability in precision gearing systems. In the case of brushless motors, bearing life is the limiting factor, so that such motors can achieve lifetimes of 20 000 h, or more, versus the 300 to 5000 h that is typical for brush-type motors [143]. For internal applications hydrodynamic or magnetic bearings are used because this type of bearing secures the longest life. In the case of blood pump, a hydrodynamically levitated impeller floats hydraulically into the top contact position. This position prevents thrombus, i.e., blood clot formation, by creating a washout effect at the bottom bearing area, a common stagnant region.

Incorporation of advanced plastic and composite components in motor and gearhead systems reduces cost, mass, and audible noise, and provides uniform products with short lead times [143].

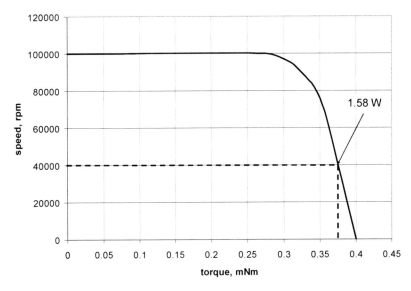

Fig. 6.2. Steady state speed–torque characteristic of a 6-mm diameter, 1.58 W PMBM manufactured by *Faulhaber*, Schonaich, Germany.

Using hardened steels for shafts and cutting gears more precisely improves gear motor capabilities by decreasing backlash and lengthening gear life.

6.3 Control

In open loop control, when power is applied to the motor, it performs some turning, running, or incremental motion, disregarding the reference position. Motors for medical devices mostly require closed loop control. For example, in a typical surgical device, such as drill or saw, the feedback that controls the motor speed may be a physician who is applying physical pressure. To provide greater consistency, or to allow a machine to perform a procedure inaccessible to human hands, it is desirable to integrate a feedback device into the instrument.

Similar to integrated electromechanical drives, in distributed control systems the entire servo system, (i.e., motor, gearhead, feedback device and microprocessor) is situated at the point where the work is done and is connected to a host system by a few wires. Eliminating multiple wires in traditional control systems, the induced noise in the system, wiring costs and complexity are reduced by a factor of three to four.

6.4 Implanted blood pumps

A *left ventricular assist device* (LVAD) is an electromechanical pump implanted inside the body and intended to assist a weak heart that cannot efficiently pump blood on its own. It is used by end-stage heart failure patients who are unable to receive a heart transplant due to donor availability, eligibility, or other factors.

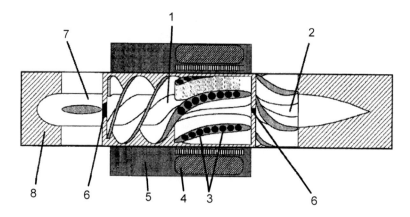

Fig. 6.3. Longitudinal section of DeBakey axial flow rotary blood pump: 1 — flow inducer/impeller, 2 — flow difuser, 3 — PMs, 4 — stator winding, 5 — stator, 6 — bearing, 7 — flow straightener, 8 — flow housing.

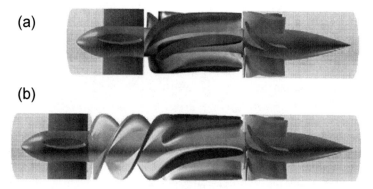

Fig. 6.4. DeBakey rotary pump: (a) original design; (b) with modifications added by NASA researchers.

6.4 Implanted blood pumps 141

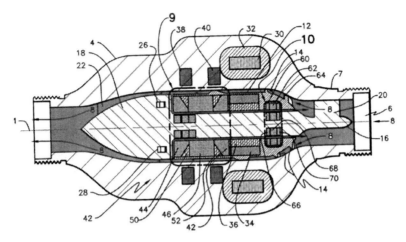

Fig. 6.5. Longitudinal section of *Streamliner* axial rotary blood pump with magnetic bearings according to US Patent No. 6244835 [11]. Description in the text.

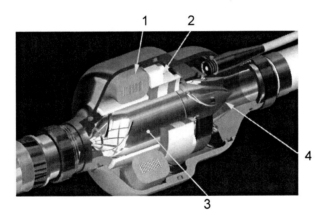

Fig. 6.6. Computer generated 3D image of *Streamliner* blood pump. 1 — stator winding of PM motor, 2 — coils of magnetic bearings, 3 — impeller, 4 — outflow hub. Courtesy of J.F. Antaki, University of Pittsburgh, PA, and *LaunchPoint Technologies*, Goleta, CA, U.S.A.

Motor driven pumps implanted in the human body must be free of shaft seals. This problem can be solved by embedding PMs in the pump rotor placed in a special enclosure and driven directly by the stator magnetic field. In this case the nonmagnetic air gap is large and high energy PMs are required.

Electromagnetic pumps for LVADs can be classified into three categories:

- 1st generation (1G), i.e., electromagnetic pulse pumps;
- 2nd generation (2G), i.e., electromagnetic rotary pumps;

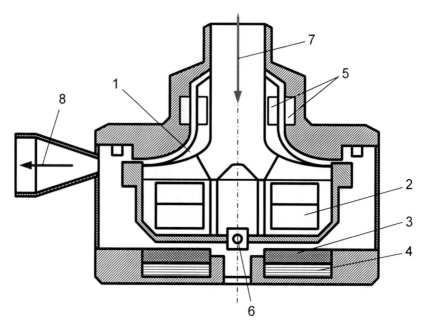

Fig. 6.7. DuraHeart® centrifugal rotary pump with axial flux PM brushless motor and magnetic levitation bearings: 1 — impeller, 2 — PM, 3 — stator winding, 4 — stator core, 5 — magnetic suspension, 6 — ceramic pivot, 7 — inlet, 8 — outlet.

- 3rd generation (3G), i.e., electromagnetic pumps with magnetic or hydrodynamic bearings.

Electromagnetic 1G pumps were driven by electromagnets, linear oscillating motors or linear short-stroke actuators. The pump, integrated with a linear actuator, is heavy, large and noisy.

DeBakey 2G LVAD axial flow rotary pump manufactured by *MicroMed Technology, Inc.*, Houston, TX, U.S.A., is driven by a novel PM brushless motor with magnets embedded in the blades of the impeller (Fig. 6.3). The inducer/impeller has six blades with eight PMs sealed in each blade and spins between 8000 and 12 000 rpm [62]. This allows it to pump up to 10 liters of blood per minute. All parts are enclosed in a sealed titanium tube. The bearings are blood-immersed pivot bearings. The titanium inlet tube attaches to the left ventricle. The outlet tube is sewn to the aorta.

An inducer added by NASA researchers eliminates the dangerous back flow of blood by increasing pressure and making flow more continuous. The device is subjected to the highest pressure around the blade tips (Fig. 6.4).

An example of 3G axial rotary blood pump is the *Streamliner*, developed at University of Pittsburgh, PA, U.S.A. [12]. The design objective was to

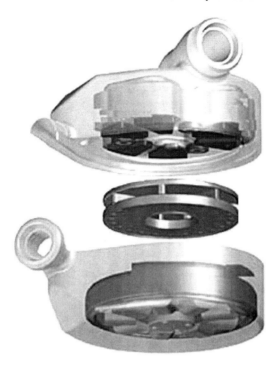

Fig. 6.8. Exploded view of DuraHeart® centrifugal rotary pump. Courtesy of *Terumo Corporation*, Tokyo, Japan.

magnetically levitate and rotate a pump impeller in the bloodstream while minimizing pump size, blood damage, battery size, and system weight.

The *Streamliner* topology is shown in Fig. 6.5 [11]. The stator (32) has a slotless core and toroidal winding. PMs (34) constitute the rotor excitation system. The key design elements are a cylindrical magnetically levitated rotating impeller (12), which is supported on PM radial bearings (9 and 10). The inner races of these bearings are fixed and supported by the outflow hub (18) and the inlet stator blades (20). The axial position of the impeller is actuated by the coils (38 and 40) interacting with the outer race magnets of the bearing (9). Sensing of the axial position is accomplished with eddy-current sensor probes (26 and 28). The outputs of these sensors are summed to render pitching motions of the impeller unobservable and decoupled from the axial feedback loop. Although the rotor is magnetically controlled in six degrees of freedom (DOF), only two DOF are actively controlled: axial and rotational motion. A 3D image of the *Streamliner* is shown in Fig. 6.6.

The DuraHeart® 3G LVAD developed by *Terumo Corporation*, Tokyo, Japan, combines centrifugal pump with magnetic levitation technologies (Figs 6.7 and 6.8). Magnetic levitation allows the impeller to be suspended within the blood chamber by electromagnets and position sensors. The three-phase,

144   6 Electric motors for medical and clinical applications

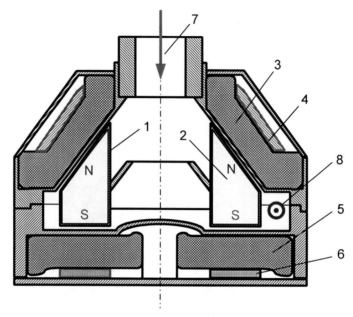

Fig. 6.9. Longitudinal section of VentrAssist^{TM} hydronamically levitated centrifugal blood pump: 1 — impeller, 2 — PM, 3 — body coil, 4 — body yoke, 5 — cover coil, 6 — cover yoke, 7 — inlet, 8 — outlet.

8-pole, axial flux PM brushless motor with slotless stator resembles a floppy disk drive spindle motor (Fig. 6.7). NdFeB PMs are integrated with the impeller. The output power of the motor is 4.5 W, speed 2000 rpm and torque 0.0215 Nm [157].

An axial flux slotless motor integrated with a centrifugal blood pump is shown in Figs 6.9 and 6.10. The so-called VentrAssist^{TM} manufactured now by Australian company *Ventracor* is a new cardiac LVAD which has only one moving part — a hydrodynamically suspended impeller integrated with a PM rotor. The hydrodynamic forces act on tapered edges of the four blades. The stator of the brushless electric motor is of slotless type and has only upper and lower coils. The three coil winding and four pole rotor use the second harmonic of the magnetic field wave to produce the torque. To provide redundancy, body coils and cover coils are connected in parallel, so that the motor still can run even if one coil is damaged.

The housing and impeller shell are made of titanium alloy Ti-6Al-4V. Vacodym 510 HR NdFEB PMs ($B_r \approx 1.4$ T, $H_c \approx 950$ kA/m) are embedded into impeller [158]. To reduce the reluctance for the magnetic flux, laminated silicon steel return paths (yokes) are designed.

The 2D FEM simulation of the magnetic field distribution is shown in Fig. 6.11. The measured performance characteristics are shown in Fig. 6.12 [157]. For output power between 3 and 7 W and speed between 2000 and 2500

6.4 Implanted blood pumps   145

Fig. 6.10. Computer generated 3D image of VentrAssist™ centrifugal blood pump. Courtesy of *Ventracor*, Australia.

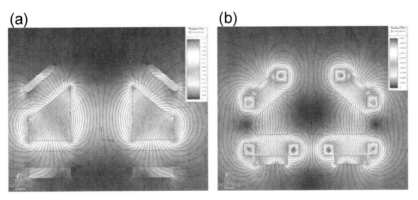

Fig. 6.11. Magnetic flux (contour plot) and magnetic flux density (shaded plot) in VentrAssist™ centrifugal blood pump excited by: (a) NdFeB PMs at no load; (b)armature (stator) coils.

rpm, the efficiency is from 45 to 48% (Fig. 6.12). At 3 W and 2250 rpm, the winding losses are 1.7 W, eddy current losses in titanium 1.0 W and losses in laminated yokes are 0.7 W [157]. At load torque 0.03 Nm the fundamental phase current is 0.72 A (Fig. 6.12).

The device weighs 298 g and measures 60 mm in diameter, making it suitable for both adults and children.

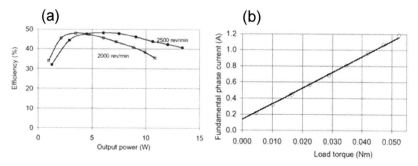

Fig. 6.12. Performance characteristics of VentrAssist^{TM}BA2-4 pump driven by a six-step sensorless inverter: (a) efficiency; (b) fundamental phase current calculated on the basis of EMF (solid line) and obtained from laboratory tests (circles) [157].

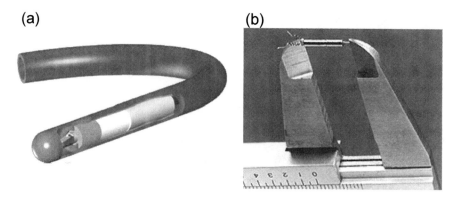

Fig. 6.13. Ultrasound motorized catheter: (a) general view; (b) 1.9-mm diameter PM brushless motor (Table 6.1). Courtesy of *Faulhaber*, Schonaich, Germany [194].

6.5 Motorized catheters

A brushless motor with planetary gearhead and outer diameter below 2 mm has many potential applications such as *motorized catheters*[2] (Fig. 6.13a), minimally invasive surgical devices, implantable drug-delivery systems and artificial organs [194]. An ultrasound catheter shown in Fig. 6.13a consists of a catheter head with an ultrasound transducer on the motor/gearhead unit and a catheter tube for the power supply and data wires. The site to be examined can be reached via cavities like arteries or the urethra[3]. The supply of power and data to and from the transmit/receive head is provided via slip rings.

---

[2] Catheter is a tube that can be inserted into a body cavity, duct or vessel.

[3] Urethra is a tube which connects the urinary bladder to the outside of the body.

The stator of the brushless motor is a coreless type with skewed winding. The outer diameter of the motor is 1.9 mm, length of motor alone is 5.5 mm and together with gearhead is 9.6 mm (Figs 6.2, 6.13b, Table 6.1. The rotor has a 2-pole NdFeB PMs on a continuous spindle. The maximum output power is 0.13 W, no-load speed 100 000 rpm, maximum current 0.2 A (thermal limit) and maximum torque 0.012 mNm (Table 6.1) [194]. The high-precision rotary speed setting allows analysis of the received ultrasound echoes to create a complex ultrasound image.

6.6 Plaque excision

The *SilverHawk*^{TM} plaque excision system is a catheter-type device for the treatment of *de novo* and restenotic lesions (abnormal structural change in the body) in the peripheral arteries [148]. Plaque excision is a minimally invasive procedure performed through a tiny puncture site in the leg or arm. The *SilverHawk*^{TM} system uses a tiny rotating blade to shave away plaque from inside the artery.

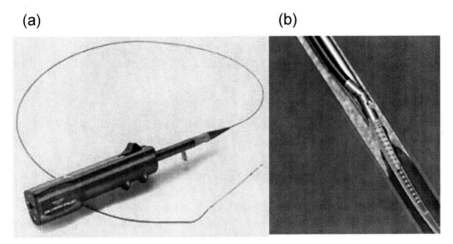

Fig. 6.14. *SilverHawk*^{TM} plaque excision system: (a) battery and motor pack with catheter; (b) shaving plaque along the vessel. Photo courtesy of *ev3 Inc.*, Plymouth, MN, U.S.A.

The *SilverHawk*^{TM} plaque excision device consists of a long catheter with a battery and motor pack on one end, and a nosecone with a carbide cutter on the other (Fig. 6.14). Depending on the vessel size, the diameter of catheter is from 1.9 to 2.7 mm and the tip length from 22 to 60 mm. The cutter (carbide blade) operates at 8000 rpm. It is 3.5 times stronger than stainless steel and 23 times stronger than calcium [65]. After delivering the catheter along a guide wire to the proximal end of the target lesion, the cutter is turned on. The cut length is determined by the operator. The plaque is shaved continuously along the vessel (Fig. 6.14b). It enables efficient treatment of long blockages within the arteries. After excision of the plaque, the cutting blade extends through the collection chamber to pack the tissue into the nosecone. The cutter is then switched off and retracted to prepare to approach the blockage from a different angle, or to treat another blockage. As it is excised, the plaque collects in the tip of the device and then is removed from the patient. All of these maneuvers are observed under fluoroscopy[4] [65].

6.7 Capsule endoscopy

Capsule endoscopy helps doctors to evaluate the condition of the small intestine. This part of the bowel cannot be reached by traditional upper *endoscopy* [5] or by *colonoscopy* [6] The most common reason for doing capsule endoscopy is to search for a cause of bleeding from the small intestine. It may also be useful for detecting polyps, inflammatory bowel disease (Crohn's disease), ulcers, and tumors of the small intestine.

Fig. 6.15. Capsule endoscope (*Pillcam*): 1 — optical dome, 2 —lens holder, 3 — lens, 4 — illuminating LEDs, 5 — complementary metal oxide semiconductor (CMOS) imager , 6 — battery, 7 — application specific IC transmitter, 8 — antenna.

---

[4] Fluoroscopy is a special type of x-ray used to project live images onto a monitor.
[5] Endoscopy is the examination and inspection of the interior of body organs, joints or cavities through an endoscope. An endoscope is a device that uses fiber optics and powerful lens systems to provide lighting and visualization of the interior of a joint.
[6] Colonoscopy is a procedure that enables a gastroenterologist to evaluate the appearance of the inside of the colon (large bowel) by inserting a flexible tube with a camera into the rectum and through the colon.

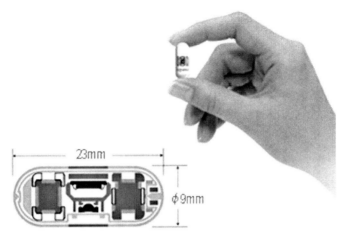

Fig. 6.16. Next generation Sayaka capsule endoscope with a stepper motor [160].

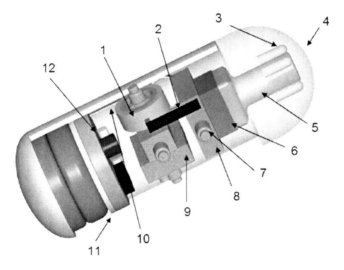

Fig. 6.17. Intracorporeal video probe IVP2. 1 — Q-PEM motor, 2 — transmission shaft, 3 — LEDs, 4 — transparent cover, 5 — optics, 6 — CMOS sensor, 7 — fixation point, 8 — camera chip, 9 — localization chip, 10 — electrical wires, 11 — batteries, 12 — data transmission chip. Courtesy of Scuola Superiore Sant'Anna, Pisa, Italy.

Approximately the size of a large vitamin, the *capsule endoscope* includes a miniature color video camera, a light, a battery and transmitter (Fig. 6.15). Images captured by the video camera are transmitted to a number of sensors attached to the patient's torso and recorded digitally on a recording device similar to a walkman or beeper that is worn around the patient's waist.

150   6 Electric motors for medical and clinical applications

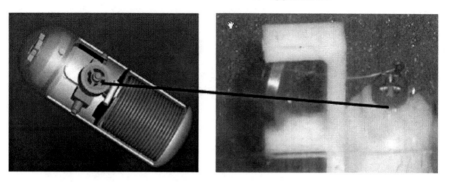

Fig. 6.18. Camera tilting mechanism with wobble motor [13].

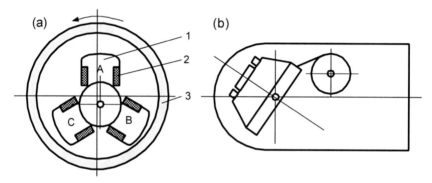

Fig. 6.19. Wobble motor: (a) cross section; (b) cam mechanism. 1 — stator core, 2 — stator coil, 3 — PM rotor wheel.

In next generation capsule endoscopy, e.g., *Sayaka capsule endoscope* [160], a *tiny stepper motor* rotates the camera as the capsule passes through the digestive tract, allowing to capture images from every angle (Fig. 6.16). Sayaka capsule is characterized by a double structure made up of an outer and an inner capsule. Whereas the outer capsule traverses through the gastrointestinal tract, the inner capsule alone spins. The spinning is produced by a small PM stepping motor with a stepping angle of 7.5^0. This stepping rotation is necessary to prevent fluctuation or blurring in the images. An eight hour, 8 m passage from entrance to exit will yield 870 000 photos, which are then combined by software to produce a high-resolution image.

The autonomous *intracorporeal video probe* (IVP) system contains a CMOS image sensor with camera, optics and illumination, transceiver, system control with image data compression unit and a power supply [13]. The optical part is located on a tiltable plate, which is driven by a *wobble motor* (Fig. 6.17). The basic concept is to use a frontal view system with a vision angle upper to 120^0 and a tilting mechanism able to steer the vision system (optics, illumination and image sensor) between about $\pm 30^0$ in one plane (Fig. 6.18).

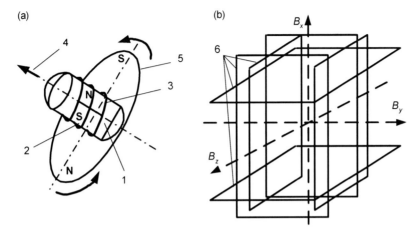

Fig. 6.20. *Olympus* capsule guided and controlled by external magnetic field: (a) capsule; (b) three pairs of external coils. 1 — capsule, 2 — PM, 3 — spiral, 4 — direction of motion, 5 — external rotating magnetic field, 6 — external coils.

By exploiting this technique, the device will perform an optimal view between ±90 degrees in the xy plane. The tilting mechanism consists of a wobble motor (Fig. 6.19a) and simple mechanical parts, such as one cam and one shaft fixed to the vision system (Int. Patent Publication WO2006/105932). The cam system transforms the rotational action of the motor in a linear action to the shaft (Fig. 6.19b). The so-called Q-PEM stepping motor can be controlled with a precision of 340 steps per revolution. The motor with outer diameter of 4 mm and thickness of 3 mm draws about 100 mW of power.

Olympus Medical Systems, Tokyo, Japan, in collaboration with Tohoku University, proposes a capsule guided and controlled by external magnetic field (Fig. 6.20). A uniform magnetic field is created by an external magnetic field generator using three pairs of opposing electromagnets arranged in three directions x, y and z. The capsule endoscope can then be turned in any desired direction by means of its built-in magnet. A rotating magnetic field can be produced that rotates the capsule, generating thrust through the spiral structure on the capsule's exterior. Coils located outside the body provide inductive power transfer to the receiving coils inside the capsule.

6.8 Minimally invasive surgery

Minimally invasive surgery (MIS), also called *laparoscopic surgery* is a modern surgical technique in which operations in the abdomen are performed through small incisions, typically from 5 to 15 mm. Traditional surgery needs much larger incisions, about 20 cm. A *laparoscope*, used in abdominal surgery, is the key instrument which consists of a telescopic rod lens system, that is usually connected to a video camera. A 'cold' light source (halogen or xenon) is used to illuminate the operative field, inserted through a 5 mm or 10 mm small tube (cannula) and connected to a fiber optic cable system to view the operative field. There are many instruments available for use in robotic surgery, including forceps, various types of grippers, scissors, electrocautery devices, staplers, needle holders, and loops for stitch threads. Instruments have four degrees of freedom (DOF) in addition to their internal freedoms (for example, opening and closing a gripper). Typical requirements are: maximum overall diameter 5 mm , minimum gripping force 5 N, minimum stroke length 1 mm, maximum gripper closing time 2 s. To create the working and viewing space, the abdominal wall is usually filled in with carbon dioxide CO_2 gas and elevated above the internal organs like a dome.

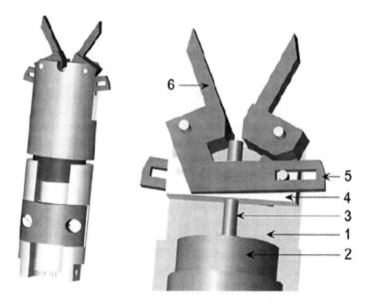

Fig. 6.21. Motorized gripper with a rotary motor: 1 — gripper frame, 2 — 6-mm diameter, 1.56-W *Faulhaber* motor (type 006B, Table 6.1) coupled with a harmonic drive 1:500 gearbox, 3 — warm, 4 — gear, 5 — arm lever, 6 — gripper [164].

End-effectors are the tools attached to the end of the robot arm that enable it to do useful work. Grippers are the most common end-effectors. Motorized lightweight grippers of laparoscopic surgery robots and surgical robotic systems can use [103, 134, 164, 185]:

- small rotary electric motors, especially PM brushless motors;
- linear motors;
- voice coil actuators (VCA);

- ultrasonic motors capable of generating a 3 DOF motion;
- magnetic shape memory actuators (MSM);
- magnetostrictive actuators;
- piezoelectric actuators
- thermal expansion actuators;
- pneumatic actuators.

A motorized gripper with a rotary motor for surgical robot is shown in Fig. 6.21 [164]. A 13-mm PM brushless motor for surgical instruments is shown in Fig. 6.1 and its specifications are listed in Table 6.2. Electromagnetic grippers can also be activated by the rotary VCA [134].

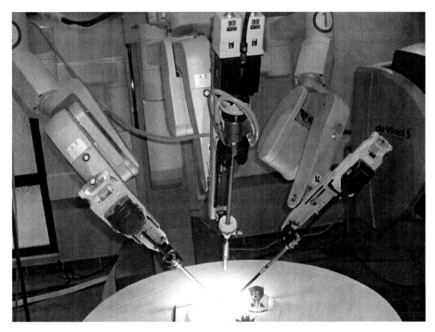

Fig. 6.22. Arms of *da Vinci*® minimally invasive surgical robot. Author's photograph.

The *da Vinci*® system, debuted in 1992, is the best known surgical robot (Fig. 6.22). The *da Vinci*® system is a master–slave sophisticated robotic platform designed to expand the surgeon's capabilities. It offers a MIS option for major surgeries. The *da Vinci*® system cannot be programmed, nor can it make decisions on its own. Every surgical maneuver is performed with direct input from the surgeon. Surgeon uses open surgery hand movements which are precisely replicated in the operative field by the instruments. Small incisions are used to introduce miniaturized wristed instruments and a high-definition 3D camera. The device includes a master, a computer controller and three

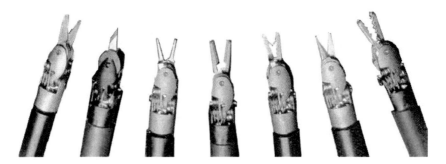

Fig. 6.23. EndoWrist® surgery tools of *da Vinci*® MIS robot. Photo courtesy of *Intuitive Surgical*, Sunnyvale, CA, U.S.A.

(a) (b)

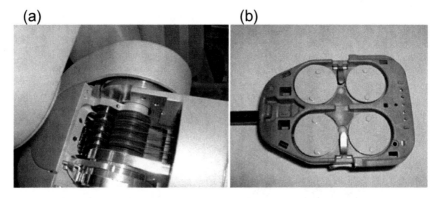

Fig. 6.24. Power transmission mechanism of *da Vinci*® surgical robot: (a) cables and pulleys in the robot's arm; (b) contact surface of the end effector with the arm. Author's photograph.

robotic arms (Fig. 6.22). The first two arms, controlled by the surgeon's left and right hands, hold the EndoWrist® instruments (grippers). The third arm positions the laparoscope (source of light and camera). The charge-coupled device (CCD) camera equipped with dual endoscope enables stereo vision. Recently, a fourth arm with EndoWrist® instruments has been added to further enhance the dexterity and autonomy of the robot. The *da Vinci*® surgery arm has 6 DOFs and 1 DOF for the tool actuation. There are two 6 DOF slave EndoWrist® manipulators, but later, an additional 7 DOF decoupled joint EndoWrist® has been added. State–of–the–art robotic and computer technologies scale, filter and translate the surgeon's hand movements into precise micro-movements of the *da Vinci*® instruments. The surgeon operates the robot and views a magnified, high-resolution 3D image of the surgical site seating at the *da Vinci*® console. The 3D vision system and camera is fully controlled by the surgeon, with the help of simple voice commands.

The *da Vinci*® was the first teleoperated medical robot to receive the U.S. Food and Drug Administration (FDA) approval in 2000 for laparoscopic radical prostatectomy (surgical removal of all or part of the prostate gland), and since then, it has been verified for other surgical procedures. From 2000 to 2006, approximately 60 000 surgeries have been performed with more than 1000 *da Vinci*® robots only in the U.S. The second generation of the *da Vinci*® robot was completed by 2003 with high definition (HD) cameras, augmented ergonomic features and a fourth robotic arm for servicing tasks. The *da Vinci*® surgery tools (grippers) are shown in Fig. 6.23.

The *da Vinci*® system offers the following benefits to patients:

- significantly less pain;
- less blood loss;
- fewer complications;
- less scarring;
- shorter hospital stay;
- faster return to normal daily activities.

The local actuation of the end-effector is a difficult challenge as a small size actuator must develop high gripping force and not exceed the temperature of human body. Also, it is hard to design a multi DOF wrist for orienting the end-effector of a 5 mm diameter robotic instrument. This is why the *da Vinci*® end effectors are not driven and controlled by locally installed electric motors or actuators, but by tendon-type actuation mechanism as shown in Fig. 6.24. The motors are placed outside end-effectors and a power transmission mechanism with cables and pulleys provide 7 DOFs to the surgical instruments.

In many diagnoses or surgeries, e.g., cancer biopsies and therapies guided by magnetic resonance imaging (MRI), tools or robotics systems require motors without metal parts and conductors carrying electric current. The motor must be constructed of nonmagnetic and dielectric materials (plastics, ceramics, rubbers, etc.) and be electricity free. The motor developed at the Johns Hopkins Urology Robotics Lab is driven by compressed air with fibre optic technology used for communications, so that all electric components are located away from the MRI scanner [186]. The new motor, dubbed *PneuStep*, consists of three pistons connected to a series of gears. The gears are turned by air flow, which is in turn controlled by a computer located in a room adjacent to the MRI machine. A precise and smooth motion of the motor as fine as 50 μm, finer than a human hair, can be achieved. Six motors are used in a surgical robot designed to carry out precise MRI-guided surgical procedures. The robot goes alongside the patient in the MRI scanner and is controlled remotely by observing the images on the MRI [186]. An MRI compatible manipulator can also employ a spherical ultrasonic motor (SUSM) [116].

6.9 Challenges

Today, therapy, surgery and health care are increasingly dependent on electrical and electronics engineering. Many medical devices require miniature and high power/torque/force density electric motors and actuators. In terms of miniaturization, high power density, high efficiency, low heat dissipation, reliability, lifetime and fault tolerance, the best motors for medical applications are PM brushless motors. Modern PM brushless motors for LVADs and other surgical devices are characterized by:

- integration of PM rotor with rotating parts of clinical devices or surgical grippers;
- large air gap in some applications, e.g., implantable blood pumps requiring high energy density PMs;
- slotless or coreless stator design (no cogging effect, no magnetic origin vibration);
- reduced winding, core and eddy current losses to keep the temperature of the motor parts below the temperature of human blood, i.e., $+36.8^{o}C$;
- application of high energy density rare earth PMs;
- reliable protection of PMs against corrosion;
- application of concentrated non-overlapping coils (Section 3.3) to the stator winding.

Torque/force density of PM electric motors/actuators is especially very important in grippers for surgical robots. So far, power transmission mechanisms outpace electric motors and actuators mounted locally in end-effectors. Moreover, electric motors and actuators cannot drive end-effectors (grippers) operating in the presence of external electromagnetic fields as, for example, MRI systems.

Small size, high torque or force, reliability, long lifetime, fault tolerance, low operating temperature and resistance to hostile environment are the most important design problems.

7

Generators for portable power applications

The energy density of currently available lithium-ion batteries is only up to 0.5 MJ/kg, their autonomy is limited and charging is a problem. On the other hand, fuel based power generation offers the energy density of about 45 MJ/kg with easy refuelling. Assuming the energy efficiency as low as 2%, engine driven miniature generators can provide a better solution for some portable electrical power supplies than batteries [184].

Miniaturisation of gas turbines causes technically difficult problems due to high rotational speeds (from 50 000 up to 500 000 rpm) and very high elevated gas temperatures (up to 1500^0 C for military engines). Scaling down gas turbine systems unfavourably influences the flow and combustion process. Their fabrication requires new high temperature materials, e.g., silicon nitride Si_3N_4 and silicon carbide SiC and 3D micromanufacturing processes [120].

7.1 Miniature rotary generators

7.1.1 Mini generators for soldiers at battlefields and unmanned vehicles

Future battlefield operations will require portable power supplies for weapons and surveillance as well as airconditioned uniforms for soldiers. Battery packs carried by soldiers are heavy and require recharging. Mini generators driven by microturbines or mini combustion engines can help to reduce the weight of batteries and recharge them everywhere using, e.g., kerosene. The power range is from 10 to 100 W for *microgenerators* (soldiers) to 100 to 1000 W for *minigenerators* (unmanned ground vehicles).

Fig. 7.1 shows a 20 to 80 W microgenerator developed by *D-star Engineering Corporation*, Shelton, CT, U.S.A [52]. The system consists of a microdiesel engine, starter/generator, power electronics, air blower, dust separator, air cleaner, catalyzed exhaust muffler and other accessories [52]. The miniature PM brushless starter/generator and solid state converter can produce 14

158   7 Generators for portable power applications

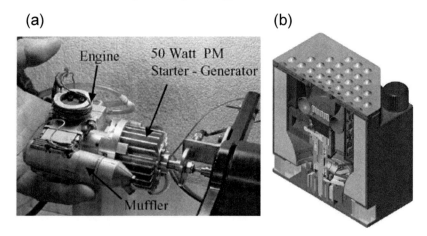

Fig. 7.1. Microgenerator pack for soldiers rated at 80 W: (a) micro engine driven PM generator; (b) generator pack [52]. Photo courtesy *D-star Engineering Corporation*, Shelton, CT, U.S.A.

V/28 V d.c. energy for charging 12/24 V batteries. The basic dimensions of pack are $127 \times 127 \times 127$ mm including a wrap-around fuel tank. The weight is

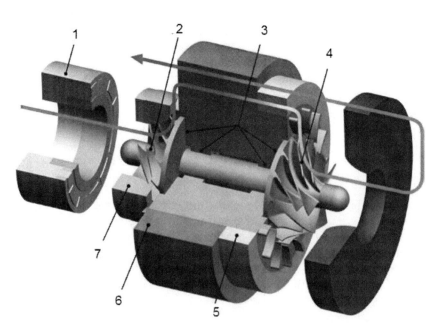

Fig. 7.2. Micro gas turbine layout. 1 — stator of generator, 2 — compressor, 3 — air bearings, 4 — turbine, 5 — combustion chamber, 6 — regenerator, 7 — rotor of generator. Courtesy of *Power MEMS*, Leuven, Belgium, [120].

projected to be 1 kg empty, and about 2 kg with a full internal fuel tank, for 24 to 36 hours of 50 W power delivery [52]. Detrimental effects are reduced in the following way [52]:

- smoke emissions are minimized by catalytic clean up of exhaust gases;
- infrared radiation (IR) is minimized by premixing the core hot air flow with the cold-side airflow prior to exit from the generator box;
- sound emissions are minimized by an inlet air filter and an exhaust catalyst/muffler, combined with wrapping the noise sources in insulation, a wrap-around fuel tank and a pouch cell battery.

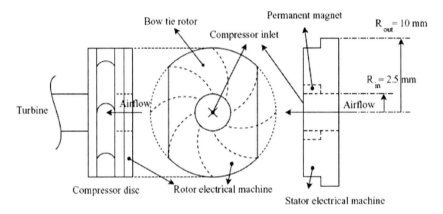

Fig. 7.3. Integration of compressor and generator/starter [184].

The micro gas turbine proposed by *Power MEMS*, Leuven, Belgium is in the centimeter range and predicted to produce electric power output of about 100 W [120]. The system basically consists of a compressor, regenerator, combustion chamber, turbine and electrical generator (Fig. 7.2). The target operation speed of this new machine is 500 000 rpm [184].

A hybrid reluctance – PM brushless machine has been selected as the generator/starter. Owing to high centrifugal stresses, a solid rotor structure is the best solution. A compact construction and minimum number of bearings (reduction of rotational losses) has been achieved by integration of the generator rotor with compressor disc. To avoid damage resulting from high stresses, PMs and coils are placed on the stator [120]. The required magnetic flux can be produced either by coils or PMs (hybrid design). Although coils lead to additional energy consumption, their use is justified in temperature environments over 130^0 C, as PMs lose their magnetic properties. The temperature sensitivity of PMs can be overcome by placing PMs on the lower temperature side of the turbine (the compressor side) and extra cooling by passing the inlet air over the PM, as sketched in Fig. 7.3 [120].

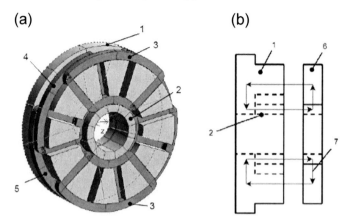

Fig. 7.4. Construction of generator/motor for *Power MEMS* micro gas turbine: (a) 3D layout; (b) magnetic flux paths. 1 — stator, 2 — PM, 3 — phase 1, 4 — phase 2, 5 — phase n, 6 — rotor, 7 — magnetic flux [184].

The PM is ring shaped and axially magnetized (Fig. 7.4a). Magnetic flux paths are shown in Fig. 7.4b. The PM flux crosses the air gap and enters the rotor. Then it goes towards the outer diameter where again it crosses the air gap towards the stator teeth. The stator back iron creates a return path for the magnetic flux.

Fig. 7.5. 1 kW, 452 000 rpm mini generator: (a) ironless stator; (b) PM rotor. Photo courtesy of *Calnetix*, CA, U.S.A.

Calnetix, CA, U.S.A. has developed a 1 kW, 452 000 rpm PM brushless minigenerator (Fig. 7.5) for U.S. Air Forces (USAF). This is an ironless design where the stator is flooded with oil to intensify cooling and increase power density. The machine was originally designed to operate on ball bearings. The efficiency is about 94 %.

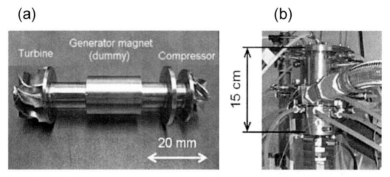

Fig. 7.6. Portable mini gas turbine system developed at Tohoku University, Japan: (a) rotor of turbine integrated with rotor of electric generator; (b) gas turbine engine. Photo courtesy of University of Tohoku, Japan.

Researchers at Tohoku University, Nano-Precision Mechanical Fabrication Lab, Japan have developed a palm-sized gas turbine engine to power autonomous robots and serve as a portable engine for personal transportation devices for elderly. The tiny engine measures 100 mm in diameter and 150 mm in length. With air bearings, 16 mm compressor rotor diameter, 17 mm turbine rotor diameter and combustion chamber, the engine can develop a rotational speed of 500 000 to 600 000 rpm (Fig. 7.6). It has been not decided yet which type off electric generator will be integrated with mini gas turbine engine.

Fig. 7.7. Micro gas turbine engines for reluctance micro starter/generators: (a) turbine engines (b) generator. Photo courtesy of University of Leicester, U.K.

University of Leicester, U.K. has developed a reluctance starter/generator for a mini gas turbine in an IEC 56 frame size. A reluctance machine enables higher speed and is more robust than an equivalent PM brushless machine. According to the University of Leicester, the controllability of SR machine is simpler than that of most inverter fed a.c. machines. Fig. 7.7 shows typical starter/generators with shaft speeds of up to 80 000 rpm and output power of 3 kW. It is predicted to use such machines for micro and mini gas turbines for portable power generation and efficient CHP systems.

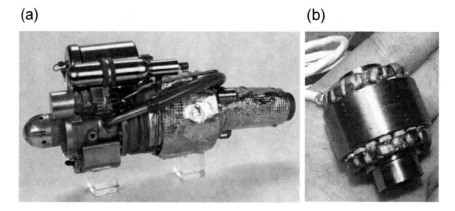

Fig. 7.8. TJ-50 miniature propulsion system: (a) prototype; (b) 1 kW, 130 000 rpm PM brushless generator [81]. Photo courtesy of *Hamilton Sundstrand Power Systems*, San Diego, CA, U.S.A.

Hamilton Sundstrand Power Systems, San Diego, CA, U.S.A. has developed a family of miniature turbojet engines to meet the needs of small, low-cost weapons system such as targets, decoys and mini cruise missiles (Fig. 7.8a) [81]. The rotational speed is 130 000 rpm and external diameter of engine is 102 mm. These engines are equipped with miniature PM brushless generators (Fig. 7.8a). The engine has an altitude start goal of 4500 m and needs to operate reliably up to 9000 m with a nominal 30 min mission time.

A 100 W, 500 000 rpm generator/starter driven by a gas turbine has been designed and constructed at Swiss Federal Institute of Technology, Zurich, Switzerland [215]. The *rms* output phase voltage is 11 V and frequency 8.4 kHz at 500 000 rpm. To increase the efficiency, a slotless stator with Litz wire windings and low loss core magnetic material have been selected. The stator core inner diameter is 11 mm, outer diameter is 16 mm and the effective length of the stator core is 15 mm. The rotor does not have any ferromagnetic core. A cylindrical Sm_2Co_{17} PM has been chosen, because the operating temperature is up to 350^0C. The outer rotor diameter with retaining sleeve is 6 mm. The stator winding resistance is 0.1 Ω and the inductance is only 3 μH [215].

A research team from the CAS Guangzhou Institute of Energy Conversion (GIEC) has recently been successful in developing a prototype for a microgenerator being just 35 mm in external diameter and 70 mm in length. With a speed of 20 000 rpm, the system can produce about 1 W of power, enough for making three LEDs shine brightly.

7.1.2 Coreless stator disc type microgenerators

Fig. 7.9 shows a disc type microgenerator developed at Georgia Institute of Technology, Atlanta, GA, U.S.A. [16, 17]. The stator uses interleaved, electroplated[1] copper windings that are dielectrically isolated from a 1-mm thick NiFeMo substrate by a 5 μm polyimide (high temperature engineering polymer) layer. The rotor consists of a 2 to 12 pole, 500 μm thick, annular PM (10 mm outer diameter, 5 mm inner diameter) and a 500 μm thick Hiperco 50 (Section 2.2) ring as a return path for the magnetic flux. SmCo PMs and Hiperco cobalt alloy have been used, because microgenerators are predicted to operate in high temperature environment, i.e., powered by gas-fueled turbine engines. Specifications are given in Table 7.1 [16].

Fig. 7.9. Experimental tests on disc type microgenerators by spinning a small magnet with the aid of air-powered drill above a mesh of coils fabricated on a chip. Photo courtesy of Georgia Institute of Technology, Atlanta, GA, U.S.A.

---

[1] Electroplating is the process of using electrical current to reduce metal cations in a solution and coat a conductive object with a thin layer of metal.

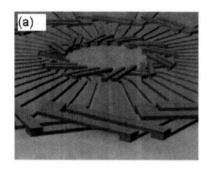 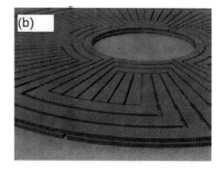

Fig. 7.10. Stator winding patterns for 8-pole microgenerator: (a) 2 turns per pole; (b) 3-turns pole [16]. Photo courtesy of Georgia Institute of Technology, Atlanta, GA, U.S.A.

Table 7.1. Specifications of a disc type microgenerator with stator electroplated winding fabricated at the School of Electrical and Computer Engineering, Georgia Institute of Technology, Atlanta, GA, U.S.A. [16, 17].

Number of poles	2 to 12
Number of turns per pole	1 to 6
Outer end turn extension	0 to 2.5 mm
Inner end turn extension	0 to 2.5 mm
Magnet outer radius, mm	5.0
Magnet inner radius, mm	2.5
Stator radial conductor outer radius, mm	4.75
Stator radial conductor inner radius, mm	2.75
Hiperco ring thickness, μm	500
Magnet thickness, μm	500
Radial conductor thickness, μm	200
End turn thickness, μm	80
Substrate thickness, μm	1000
Air gap, μm	100
Power electronics equivalent resistance, mΩ	100

A 16 W of mechanical–to–electrical power conversion and delivery of 8 W of d.c. power to a resistive load at a rotational speed of 305 krpm has been demonstrated [16]. The power density per volume of the disc type microgenerator with stator electroplated winding was 59 W/cm^3.

7.2 Energy harvesting devices

Self-powered microsystems have recently been considered as a new area of technology development. Interest in self-powered microsystems have been ad-

7.2 Energy harvesting devices   165

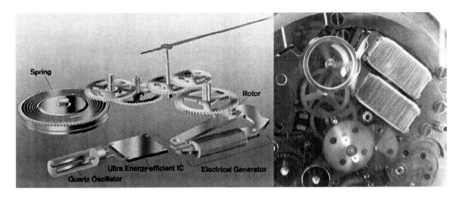

Fig. 7.11. Electro mechanism of *Seiko* kinetic watch.

Fig. 7.12. Electromechanical generator harvesting vibration energy (46 μW, 428 mV rms, 52 Hz, 1 cm^3 volume). Four PMs are arranged on an etched cantilever beam with a wound coil located within the moving magnetic field [199]. Photo courtesy of University of Southampton, U.K.

dressed in several papers, e.g., [7, 75, 112, 119, 172, 173]. Main reasons which stimulate research in this new area are:

- large numbers of distributed sensors;
- sensors located in positions where it is difficult to wire or charge batteries;
- reduction in cost of power and communication;
- Moore's law (the number of transistors per unit area of an IC doubles every 18 months).

Microsystems can be powered by energy harvested from a range of sources present in the environment. Solar cells, thermoelectric generators, kinetic generators, radio power, leakage magnetic or electric fields are just a few examples. In some applications, e.g., container security systems, condition monitoring of machine parts (motors, turbines, pumps, gearboxes), permanent embedding

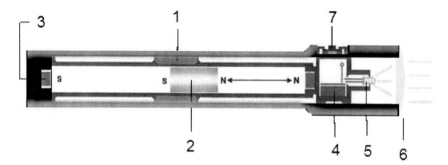

Fig. 7.13. *NightStar*® flashlight with moving magnet linear generator. 1 — stationary coil, 2 — moving magnet, 3 — repulsion magnet, 4 — capacitor and charging circuit, 5 — LED, 6 — precision acrylic lenses, 7 — sealed magnetic switch. Photo courtesy of *Applied Innovative Technologies*, Fort Lupton, CO, U.S.A.

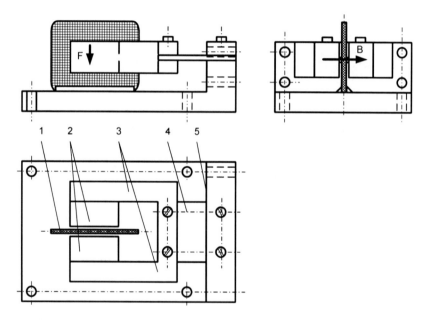

Fig. 7.14. Principle of operation of a simple electromechanical energy harvesting device. 1 — stationary coil; 2 — NdFeB PMs; 3 — mild steel magnetic circuit; 4 — cantilever beam (flat spring); 5 — base [73].

in inaccessible structures (bridges, towers, roads), or animal tracking, the only source of electrical energy is the kinetic energy.

The *harvesting of kinetic energy* is the generation of electrical power from the kinetic energy present in the environment. The nature of the kinetic energy harvesting mechanism in a self contained system depends upon the nature of

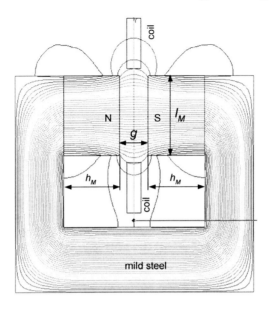

Fig. 7.15. Magnetic field distribution excited by NdFeB PMs and coil current (300 turns, 0.07 A) in an electromechanical energy harvesting device with cantilever beam [73].

the application [75, 112, 183]. Kinetic *energy harvesting devices* can be divided into two groups:

- *acceleration/vibration* and *spring mass* system devices, e.g., kinetic watches *Asulab (Swatch Group)* and *Seiko* (Fig. 7.11), cantilever beam vibration generators (Fig. 7.12), moving magnet linear generators (Fig. 7.13);
- *repeated straining physical deformation* devices, e.g., piezoelectric generators or magnetic shape memory (MSM) generators [119, 173, 183].

The *Seiko* kinetic watch (Fig. 7.11) uses the movement of the wearer's arm to produce the electrical energy to keep the watch running. The movement of wrist rotates the oscillating pendulum (weight) attached to a relatively large gear which is engaged with a very small pinion. The pendulum can spin the pinion up to 100 000 rpm. The pinion is coupled to a miniature electrical generator which charges a capacitor or a rechargeable battery. Once the kinetic watch has been fully charged, the wearer can enjoy a full six months of continuous use.

In *electromechanical energy harvesting device* with *cantilever beam* (Figs 7.12, 7.14) the input mechanical energy coming from external source of vibration (vehicle, marine vessel, engine, road, etc.) is converted into the output electrical energy. A spring–mass system with moving PM is mechanically excited by external vibration. The voltage is induced in a stationary coil that is embraced by PM poles. When the coil is loaded with an external impedance,

168   7 Generators for portable power applications

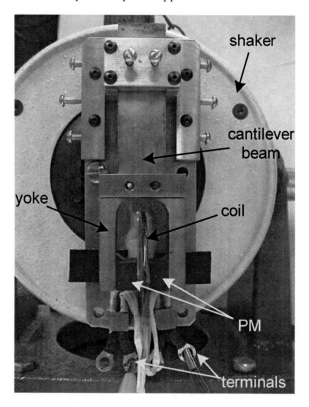

Fig. 7.16. Prototype of energy harvesting electromechanical device placed on a small variable-frequency shaker [73]. Photo courtesy of United Technologies Research Center, East Hartford, CT, U.S.A.

an electric current proportional to the induced EMF arises in the external circuit. Fundamental equations for performance calculations can been derived on the basis of elementary beam theory and circuit analysis. The magnetic field distribution excited both by PMs and coil, as obtained from the 2D FEM, is shown in Fig. 7.15.

In *NightStar®* flashlight (Fig. 7.13) the kinetic energy of motion is transformed into electrical energy by means of repeatedly passing a PM through a stationary coil. Stationary PMs oriented to repel the moving magnet are mounted at both ends of the flashlight. Repulsion PMs smoothly decelerate and accelerate the moving magnet back through the stationary coil. Kinetic energy is therefore efficiently converted into electrical energy. The generated electrical energy is stored in a capacitor. *NightStar®* is most effectively recharged when it is turned off and shaken between two and three times per second over a distance of approximately 5 cm. On a full charge *NightStar®* will provide highly effective illumination for over 20 minutes.

7.2 Energy harvesting devices

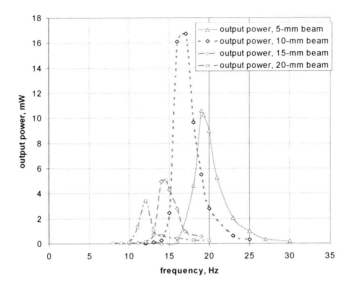

Fig. 7.17. Output power versus frequency for different lengths of the steel cantilever beam of an electromechanical energy harvesting device shown in Fig. 7.16 [73]

Portable electromechanical energy harvesting devices with the magnetic field excitation system integrated with the cantilever beam (flat spring) and a stationary multiturn coil are the most efficient devices (Figs 7.12, 7.14 and 7.16) [73]. The maximum generated energy is when the *mechanical resonance* occurs, i.e., when the natural frequency of the cantilever beam-based vibrating system is the same or close to the input frequency of vibration. The output power for different lengths of the steel cantilever beam of an electromechanical energy harvesting device shown in Fig. 7.16 is plotted versus frequency in Fig. 7.17. Potential applications of energy harvesting devices include:

- condition-based monitoring of machinery and structures;
- wireless sensors installed in security systems of containers or trailers,
- implanted medical sensors;
- wearable computers;
- intelligent environments (*smart space*), etc.

It is, in general, not difficult to design electromechanical energy harvesting devices in the range of microwatts, but it is very difficult to design properly functioning devices rated at milliwatt level. Electromechanical energy devices are down scalable to the microelectromechanical system (MEMS) levels. In this case such effects as micro-collisions, air friction dissipation, and cushioning effects cannot be ignored in their analysis and synthesis [73].

8
Superconducting electric machines

SC field excitation winding can provide high magnetic flux density in the air gap at zero excitation losses. Such field excitation systems lead to performance characteristics not achievable so far by classical field excitation systems, e.g., with copper coils or PMs. A PM motor rated at 7 MW for marine pod propulsor weighs about 120 t. An HTS synchronous motor concept would drastically reduce the weight of a podded electromechanical drive by 50%.

8.1 Low speed HTS machines

8.1.1 Applications

Low speed HTS machines can find applications in the next few years as generators rated up to 10 MW and ship propulsion motors rated up to 40 MW. Larger HTS machines at present time show small improvements in their efficiency as compared with classical synchronous generators. For example, the prototype of a 100 MVA, 60 Hz HTS synchronous generators at *General Electric* was abandoned in 2006 due to high costs of research and development in comparison with minimal efficiency increase over modern very efficient classical synchronous generators.

Various development programs of HTS motors for ship propulsion are currently under way. *American Superconductors* (AMSC) delivered to U.S. Navy in 2006 a 36.5 MW, 120 rpm synchronous motor with HTS field excitation winding cooled by liquid nitrogen [9].

Another development program is carried out in Japan by an academic-industrial consortium including the HTS conductor manufacturer *Sumitomo Electric Industries Ltd.* (SEI). This group develops a 400 kW HTS ship propulsion motor, which is cooled by liquid nitrogen. Other than the AMSC motor, the SEI concept implies a disc type machine design without rotating HTS coils [9].

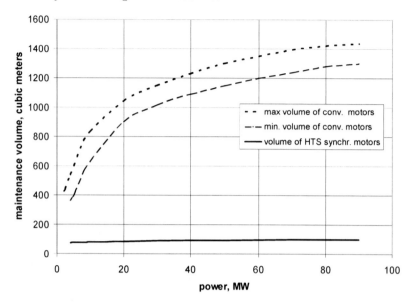

Fig. 8.1. Volume versus power of conventional and HTS synchronous motors for marine propulsion. Source: *American Superconductors* [9].

The combination of a conventional stator ferromagnetic core with a rotor employing an HTS field excitation winding and a warm or cold ferromagnetic core is the lowest power density concept of an SC machine. Cold rotors are acceptable up to about 10 MW output power. Rotors with cold ferromagnetic cores are not feasible for large power generators rated above 150 MVA, because of the tremendously long cooldown and warm-up times (larger volume of machine).

Synchronous machines with HTS rotor field excitation windings show the following general advantages:

- increase in machine efficiency beyond 99% by reducing power losses by as much as 50% over conventional generators;
- energy saving;
- reduced weight;
- reduced acoustic noise;
- lower life-cycle cost;
- enhanced grid stability;
- reduced capital cost.

By using an SC wire for the field excitation winding, the field winding losses $I_f^2 R_f$ can be practically eliminated, since the field winding resistance $R_f = 0$. The magnetic flux excited in the stator (armature) winding by the rotor excitation system is not limited by saturation magnetic flux density of the

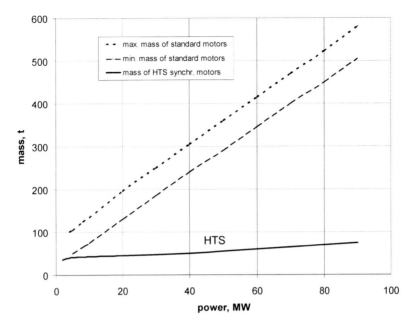

Fig. 8.2. Mass versus power of conventional and HTS synchronous motors for marine propulsion. Source: *American Superconductors* [9].

ferromagnetic core, because the stator system can be constructed without stator ferromagnetic teeth. Slotless armature system means that losses in the armature teeth do not exist and the distribution of the stator air gap magnetic flux density waveform can be sinusoidal.

8.1.2 Requirements

SC motors are much smaller than classical synchronous motors for ship propulsion. Fig. 8.1 shows comparison of volume (in cubic meters) of conventional marine propulsion motors with HTS synchronous motors, while Fig. 8.2 compares their mass.

The efficiency of HTS, classical synchronous and induction motors for ship propulsion is shown in Fig. 8.3. The efficiency curve of HTS synchronous motors is practically flat for shaft power from 5 to 100%. This is a very important advantage of synchronous motors with SC field excitation winding.

Fig. 8.4 shows the construction of a large synchronous motor with salient pole ferromagnetic rotor core and rotor HTS field excitation winding for ship propulsion [93].

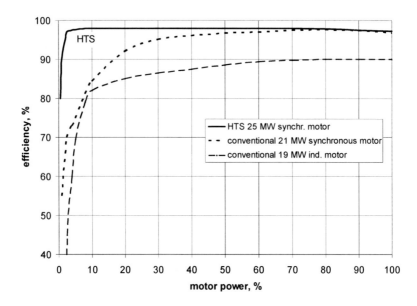

Fig. 8.3. Efficiency versus power of conventional and HTS synchronous motors for marine propulsion. Source: *American Superconductors* [9].

Table 8.1. Design parameters for the 5 MW 1G HTS synchronous motor [9].

Rated power, MW	5.0
Rated phase voltage, kV	2.4
Rated rms phase current, A	722
Power factor at rated load,	1
Rated speed, rpm,	230
Stator current frequency at rated speed, Hz	11.5
d-axis synchronous reactance, p.u.	0.32
d-axis transient reactance, p.u.	0.24
d-axis subtransient reactance, p.u.	0.16
Armature short-circuit time constant, p.u.	0.069
d-axis subtransient short-circuit time constant, s	0.02
q-axis subtransient short-circuit time constant, s	0.028

8.1.3 HTS synchronous motor for ship propulsion rated at 5 MW

Under the U.S. Navy's Office of Naval Research (ONR) program, *American Superconductors* designed, built and factory tested the 5 MW synchronous motor with 1G HTS rotor (Fig. 8.5) for ship propulsion integrated with a commercially available power electronic drive system. The rotor with BSCCO field excitation winding shown in Fig. 8.6 has salient ferromagnetic poles. The low-speed, 230 rpm, high-torque, 5 MW HTS motor was a critical development

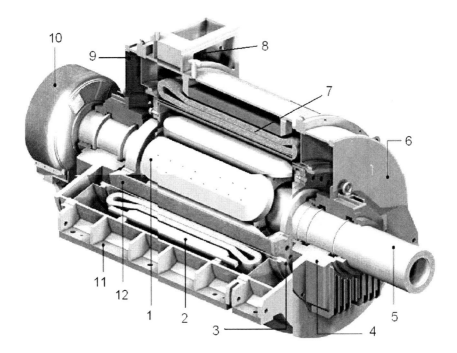

Fig. 8.4. Construction of a 5 MW synchronous motor with 1G BSCCO HTS rotor for ship propulsion. 1 — HTS rotor coil, 2 — stator yoke, 3 — electromechanical shield, 4 — bearing, 5 — drive shaft, 6 — motor enclosure, 7 — stator copper coil, 8 — power terminal box, 9 — stator coolant tank, 10 — brushless exciter, 11 — motor mount, 12 — vacuum chamber. Courtesy of *American Superconductors* [93].

milestone on the path to 25 MW and 36 MW motors, which are the power ratings expected to be utilized on electric warships and on large cruise and cargo ships. HTS motors of these power ratings are expected to be only one-fifth the volume of conventional motors.

The design specifications of the 5 MW 1G HTS synchronous motor are shown in Table 8.1. In 2003, the motor final assembly was completed. The open circuit and short circuit tests according to IEEE Standard 115 [193] were performed at *Alstom* facility in Rugby, U.K., to determine motor parameters, which matched calculated parameters quite well. A summary of the key factory test results is as follows:

- The stator design voltage of 4160 V (line-to-line) was demonstrated by open circuit testing at the rated operating speed of 230 rpm.
- The rated stator operating current of 722 A was determined and demonstrated via a short circuit testing. This test shows that the stator cooling complies with design requirements.

176   8 Superconducting electric machines

Fig. 8.5. General view of the 5 MW synchronous motor with HTS rotor for ship propulsion under tests at *Alstom* facility in Rugby, U.K. Photo courtesy of *American Superconductors*, MA, U.S.A. [9].

- The core and eddy current losses (open circuit test) were consistent with the design.
- The total losses for the full load operating conditions were determined from the open circuit and short circuit tests, and the full load efficiency was demonstrated to be greater than the contract goal of 96%.
- The functionality of the brushless exciter, controls and telemetry were demonstrated.
- The cryogenic system and HTS coils operated precisely as designed.
- Stray flux density was below 1 Gs at 1 cm away from the frame.

Electrical parameters obtained from experimental tests are close to predicted values, which confirms that the design approach was right. The performance characteristics of the *American Superconductors* 5 MW synchronous motor with HTS field excitation winding measured at *Alstom* facility [55] are plotted in Figs. 8.7, 8.8, 8.9 and 8.10.

Fig. 8.6. BSCCO HTS rotor assembly with exciter of the 5 MW synchronous motor manufactured by *American Superconductors* [9].

8.1.4 Test facility for 5 MW motors

Under U.S. Navy supervision, the 5 MW 1G HTS motor has successfully completed load and ship mission profile dynamic simulation tests that were

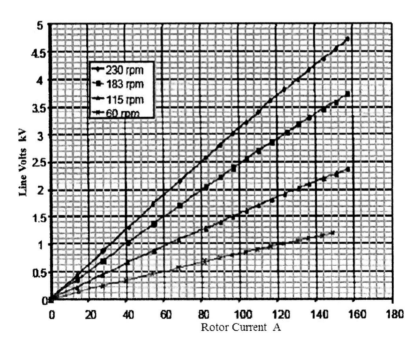

Fig. 8.7. Open circuit characteristics of the 5 MW synchronous motor with HTS field excitation winding manufactured by *American Superconductors* [9, 55].

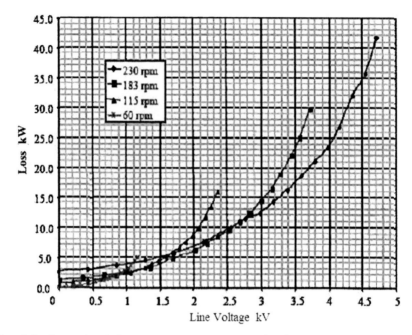

Fig. 8.8. Open circuit losses versus stator terminal voltage at constant speed of the 5 MW synchronous motor with HTS field excitation winding manufactured by *American Superconductors* [9, 55].

conducted at the Navy's Center for Advanced Power Systems (CAPS) at Florida State University in Tallahassee, FL, U.S.A. The CAPS test facility is shown in Figs 8.11 and 8.12. CAPS also repeated and confirmed the no-load as well as partial load tests according to IEEE Standard 115 [193] that had previously been conducted at the *Alstom* facility in Rugby, U.K.

The tests were designed to evaluate three U.S. Navy goals:

(a) to establish the full power capability of the motor with several longer heat or endurance runs;
(b) to establish as accurately as possible important machine parameters by a variety of methods;
(c) to investigate the dynamic performance in a simulated at–sea environment.

Successful testing of the 5 MW prototype has validated *American Superconductors* electromagnetic, mechanical and thermal analytical models and established commercial viability of HTS ship propulsion motors.

The advanced power systems test bed consists of a highly flexible, reconfigurable 4.16 kV distribution system with a stiff connection through a dedicated 7500 MVA service transformer to an adjacent utility substation 115/12.5 kV distribution transformer (Fig. 8.12). Connected into the 4.16 kV

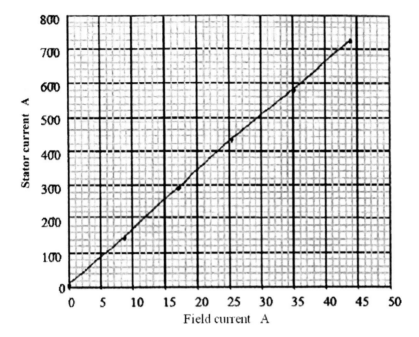

Fig. 8.9. Stator current versus field excitation current at shorted stator terminals, generating mode and rated speed of the 5 MW synchronous machine with HTS field excitation winding manufactured by *American Superconductors* [9, 55].

system is a 5 MW dynamometer comprised of two 2.5 MW GE IMs driven by *Toshiba* 4-quadrant PWM variable frequency drives (Figs. 8.11 and 8.12). A 4.16kV, 5MW VVVF converter can provide a high degree of control over a.c. and d.c. experimental buses at high power levels. The 5 MW dynamometer is housed in a high-bay test area equipped with 40 t of overhead crane capacity and a closed loop cooling water system distributed throughout the building.

CAPS operated the motor at full load 5 MW and at full speed 230 rpm for a total of 21 hours, and confirmed that the motor achieved steady state temperature both in the rotor and in the stator. The actual temperatures measured correlated closely to results predicted by AMSC and *Alsthom* for the machine. This load testing demonstrated that the HTS motor meets its specified performance and power rating under the stresses of operating conditions. An important aspect of the new results obtained at CAPS on the 5-MW motor is the validation of AMSC electromagnetic, mechanical and thermal analytical models for HTS ship propulsion motors, which is a vital step in the development cycle for advanced electrical machines.

To simulate operation of the motor in an at–sea environment up to Sea State 5 (moderately strong sea), the CAPS testing imposed an increasing scale of 0.5% to 10% in torque variations on the motor. The test results confirmed

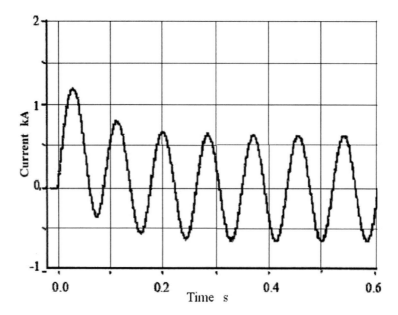

Fig. 8.10. Stator current waveform for sudden short circuit at 7% voltage of the 5 MW synchronous motor manufactured by *American Superconductors* [9, 55].

that the motor should perform as designed in representative sea states. To complete the test program, CAPS performed *hardware-in-the-loop* simulation tests. In this test phase, direct control of the motor system emulated the way the motor, drive, and entire ship's integrated power system (IPS) would respond during transients.

8.1.5 HTS motor for ship propulsion rated 36.5 MW

American Superconductors, partnered with *Northrop Grumman* assembled, tested and shipped in 2006 to US Navy the 36.5 MW HTS ship propulsion motor, under contract with the Office of Naval Research (ONR). The 36.5 MW synchronous motor shown in Fig. 8.13 has been specifically designed to power the next generation of U.S. Navy warships, e.g., future surface combatants such as the DD(X) and CG(X). The motor weighs only about 75 t versus 200 t weight of advanced IMs developing similar torque. Main features of the 36.5 MW HTS synchronous motor are:

- Application of 1G BSCCO wires for the rotor field excitation winding.
- High power density, because the HTS field winding produces magnetic fields higher than those of conventional machines resulting in smaller size and weight.

Fig. 8.11. Test facility for motors and solid state converters rated up to 5 MW at CAPS, Florida State University, FL, Tallahassee. Photo courtesy of CAPS.

- High partial load efficiency: HTS motors have higher efficiency at part load (down to 5% of full speed), that results in savings in fuel use and operating cost. The advantage in efficiency can be over 10% at low speed.

Table 8.2. Comparison of two prototypes of HTS synchronous motors for marine propulsion manufactured by *American Superconductors* [9].

Output power, MW	5.0	36.5
Speed, rpm	230	120
Torque efficiency, %	96	97
Power factor	1.0	1.0
Number of phases	3	9
Number of poles	6	16
Terminal voltage, kV	4.16	6.0
Armature current, A rms	722	1,270
Frequency, Hz	11.5	16
HTS wires	1G BSCCO	
Mass, t	23	75
Dimensions (LxWxH), m	2.5x1.9x1.9	3.4x4.6x4.1
Stator winding cooling	Liquid	
Power electronics	Commercial marine	

182   8 Superconducting electric machines

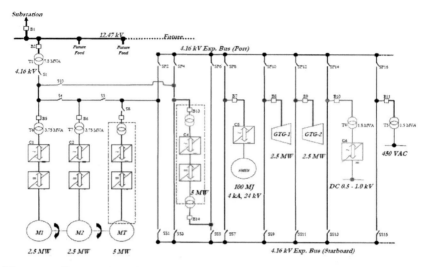

Fig. 8.12. Power circuit for testing motors and solid state converters rated up to 5 MW at CAPS, Florida State University, FL, Tallahassee. Courtesy of CAPS.

Fig. 8.13. 36.5 MW synchronous motor with 1G BSCCO HTS rotor for ship propulsion manufactured by *American Superconductors*. Photo courtsy of *American Superconductors*, Devens, MA, U.S.A.

- Low noise: HTS motors have lower sound emissions than conventional machines due to slotless stator winding.
- Low synchronous reactance: HTS air-core motors are characterized by a low synchronous reactance which results in operation at very small load

angles. Operating at a small load angle provides greater stiffness during the transient and hunting oscillations.
- HTS motors generate EMF practically free of higher harmonics.
- Cyclic load insensitivity: HTS motor field windings operate at nearly constant temperature unlike conventional motors and, therefore, are not subject to thermal fatigue.
- Maintenance: HTS motors compared to conventional motors will not require the common overhaul, rewinding or re-insulation.

Comparison of specifications of 5 MW and 36.5 MW HTS motors fabricated by AMSC is given in Table 8.2.

8.1.6 Superconducting synchronous generators

Similar to SC motors, in *SC synchronous generators* only the d.c. field excitation winding is designed as an SC winding. The stator (armature) winding is a three-phase slotless winding with liquid cooling system.

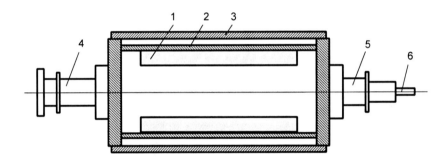

Fig. 8.14. Air core rotor of a superconducting synchronous generator [175]. 1 — HTS coil, 2 — torque tube, 3 — electromagnetic shield, 4 — turbine end shaft, 5 — collector end shaft, 6 — transfer coupling.

HTS generators are expected to be about half (50%) the size and weight of a conventional machine. These generators have a low synchronous reactance that results in operation at very small load angles. HTS generators have 50% to 70% the losses of a conventional generator and retain high efficiency down to 5% of the rated load. The cryogenic cooling system power consumption is less than a few percent of the total losses in the machine. HTS generators display the following benefits:

- higher power density than that of classical synchronous generators;
- high efficiency at all loads - down to 5% of full load;
- low noise;
- superior negative sequence capability;

- excellent transient stability;
- low synchronous reactance – small load angle;
- low harmonic content;
- increased fatigue reliability;
- require less maintenance than classical synchronous generators.

A classical SC generator has coreless rotor with SC field excitation winding (Fig. 8.14). Since very high magnetic flux density is excited in the air gap, it provides the highest torque density of all types of electric machines. The drawback is that the torque producing electromagnetic force acts directly on the SC winding, which has to be transmitted to the room temperature shaft through a low-thermal conductivity *torque tube*. This type of generator also requires the largest amount of superconductor. Practical limits on the air gap flux density result from the peak field in the coil and the amount of superconductor required. This option is the best suited rather for medium speed machines. Although it offers very high torque density, the difficulty of maintaining SC coil integrity at very high speeds makes it less attractive for high-power-density applications at high speeds [175].

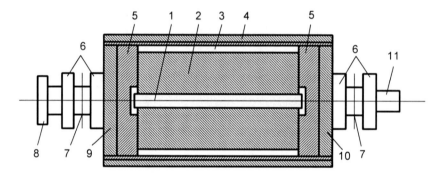

Fig. 8.15. Rotor with ferromagnetic core of an SC synchronous generator [175]. 1 — HTS coil, 2 — rotor body, 3 — vacuum enclosure, 4 — electromagnetic shield, 5 — spacer, 6 — oil seal lands, 7 — bearing centre, 8 — coupling flange, 9 — turbine end stub shaft flange, 10 — collector end stub shaft flange, 11 — collector and cryogenic connections.

In the case of rotor with ferromagnetic core (Fig. 8.15), the electromagnetic torque is taken by the steel or other ferromagnetic alloy, which can be at ambient temperature. Because the magnetic permeability of the steel is much higher than that of the air, the required excitation MMF is significantly lower and consequently less SC material is used by the rotor excitation system. In practice, a ferromagnetic core limits the air gap magnetic flux density to lower values than in the case of air core generator, but 2 T magnetic flux density (slightly lower than saturation magnetic flux density) is competitive to that obtained in classical electric machines. Thus, the torque density is sometimes not worse than that developed by air core SC generators [175]. Smaller coils and simpler mechanical support allow for increasing the speed and power density.

8.1.7 Dynamic synchronous condenser

On the basis of its own proprietary rotating machine design and hardware expertise during the last decade, *American Superconductors* has developed a *dynamic synchronous condenser* (DSC) called SuperVAR^{TM} [9]. The primary function of the DSC is to compensate the reactive power on the utility grid. Reactive power component can be either inductive (voltage leading current, corresponding to inductive reactance) or capacitive (voltage lagging current, corresponding to capacitive reactance). Reactive power does not provide any useful energy and circulates only in a.c. power grids between generators and their loads, thus contributing to increase in the *rms* current and transmission line losses. Reactive power limits the amount of real power that can be handled on a transmission or distribution grid. A synchronous condenser compensates either an inductive or a capacitive reactance by introducing the opposite type of reactance.

The DSC design offers a cost effective alternative to reactive power compensation, compared to the traditional synchronous condensers. The DSC allows the generators to operate optimally in producing real power for which a power company is ultimately paid. It is also cost-effective compared to alternatives that use capacitor banks or power electronic systems. Main features of DSC developed by *American Superconductors* are as follows:

- 10 MVAR, 13.8 kV at 60 Hz;
- steady state continuous duty;
- economic source for dynamic reactive power support;
- supports transient loads at multiples of machine ratings;
- high machine operating efficiency;
- long winding life.

8.1.8 HTS synchronous generators developed by *Siemens*

HTS synchronous machine rated at 400 kW

Starting in 1999, a four-pole HTS synchronous machine was designed and built at *Siemens*, Germany. Selected specifications are given in Table 8.3. This machine was not intended to be a prototype, but to show the feasibility of all the components that had to be designed. Starting with the rotating cryostat ranging over the torque transmission tubes and ending with the racetrack HTS coils, these components cover all the major parts of the machine. Additionally, a reliable thermosyphon-based cooling system was designed. Fig. 8.16 shows the fabricated prototype in the test bed [135, 136]. The cooling system with the cryocooler is located on the left side. The motor itself is mounted in a normal welded housing of a classical synchronous generator with a shaft height of 355 mm.

Fig. 8.16. *Siemens* 400 kW HTS synchronous machine in the test bed. Photo courtesy of *Siemens* [135, 136].

The efficiency achieved was 96.8% including the cryogenic refrigerator, as compared with 95.7% for the state–of–the art conventional machine.

Table 8.3. Specifications of the 400 kW *Siemens* model machine.

Number of phases	3
Rated frequency, Hz	50
Rated voltage, V	400
Nominal power, kW	400
Rated speed, rpm	1500
Nominal torque, N m	2600
Power overload capability	150%/15 min
Torque overload capability, %	700
Synchronous reactance, p.u.	0.15
Rotor cooling power at 25K, W	approx 25
Efficiency, %	> 96 including cryogenics

HTS synchronous generator rated at 4 MVA

Siemens has put into operation the first in the world 4 MVA HTS synchronous generator (Figs 8.17 and 8.18) to provide an extremely energy efficient electric power source for future ships. The new generator is especially suitable for cruise ships and large motor yachts. The 1G HTS wire was supplied by *European Advanced Superconductors* KG (EAS), Hanau, Germany.

Fig. 8.17. 4 MVA synchronous machine with HTS excitation winding developed by *Siemens* for ship propulsion in Large Drives System Test Center, Nuremberg, Germany. Left: 2×2.5 MW load machines under a sound reducing cover. Center: 4 MVA generator with cooling unit. Photo courtesy of *Siemens*.

The project, partly funded by the German Ministry of Education and Research (Bundesministerium fur Bildung und Forschung), was started in 2002. The first-ever synchronization of an HTS generator onto the grid was carried out in 2005 at Large Drives System Test Center, Nuremberg, Germany.

The following special construction features have been applied:

- rotor with pancake coils made out of 1G HTS tapes (Bi-2223);
- speed range up to 3600 rpm in order to allow variable speed drive operation (motor);
- medium voltage stator slotless copper winding configurable for either 6.6 kV or 3.3 kV; the latter due to planned converter operation (two terminal boxes enlarging the width of the machine by approximately 25% are provided);
- air–to–water stator cooler with forced ventilation.

The main technical specifications of the 4 MVA generator are listed in Table 8.4. It must be noted that 2-pole conventional generators with high voltage

Fig. 8.18. Cryogenic cooling system of the 4 MVA *Siemens* HTS generator. The third refrigerator is hidden behind the right one. Photo courtesy of *Siemens*.

stator winding and copper field winding are available for higher power ratings only. Therefore, a 4-pole conventional generator was used for comparison.

Table 8.4. Specifications of the 4 MW *Siemens* model machine.

Specifications	HTS generator	Conventional generator
Number of phases	3	3
Nominal power, kW	4000	4000
Nominal a.c. voltage at 60Hz, kV	6.6 (3.3)	6.6
Nominal torque, kNm	10.6	21
Nominal speed, rpm	3600	1800
Number of poles,	2	4
Synchronous reactance, p. u.	0.41	2.22
Nominal current, A	350 (700)	350
Class of stator winding insulation	F (155°C)	F (155°C)
Initial symmetrical short circuit current/nominal current ratio	16	16
Two pole short circuit torque/nominal torque ratio	7	8
Steady short circuit current/nominal current ratio	4	1.2
Underexcited operation at $cos\phi = 0$	Yes	No
Nominal cryocooler power at 25K, W	50	N/A
Required electrical power for 2 cryocoolers, kW	12	N/A
Efficiency ($cos\phi = 0$), cryocooler included, %	98.7	97.0
Length x Width x Height including cooling system, m	3.3 × 1.9 × 1.8	2.6 × 2.2 × 2.7
Foot print, m	1.9 × 1.2	1.8 × 1.8
Shaft height, m	0.5	0.8
Weight, t	7	11

8.1.9 Japanese HTS machines

HTS synchronous motor rated at 12.5 kW

A Japanese industry-academia group, called *frontier group* that includes *Sumitomo Electric Industries, Ltd.* [1] (SEI), the manufacturer of HTS wires, has developed a 12.5 kW synchronous motor with HTS BSCCO field excitation winding [187, 190]. This motor has been integrated with a pod propulsor. The pod propulsor has outer diameter of 0.8 m and is 2 m long. The diameter of the propeller is 1 m. Using a larger conventional motor, the outside diameter of the pod propulsor would be much larger.

This HTS motor is an disc type (axial flux) three phase, eight-pole machine. The stationary field excitation system and rotating armature simplify the cooling system. Technical specifications of the HTS motor are given in Table 8.5. The expanded view of the 12.5 kW HTS motor is shown in Fig. 8.19. Main features of this motor include:

---

[1] *Sumitomo Electric Industries, Ltd., Fuji Electric Systems Co., Ltd., Hitachi, Ltd., Ishikawajima-Harima Heavy Industries Co., Ltd., Nakashima Propeller Co., Ltd., Taiyo Nippon Sanso Corporation*, University of Fukui (H. Sugimoto)

190   8 Superconducting electric machines

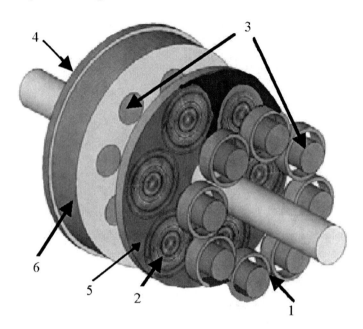

Fig. 8.19. Computer created 3D image of the disc type 12.5 kW HTS synchronous motor for pod propulsor: 1 — HTS field excitation coils, 2 — armature excitation coil, 3 — cores, 4 — back yoke, 5 — fiberglass reinforced plate of cryostat, 6 — cryostat [187, 190].

- machine uses low cost 1G BSCCO HTS wires that were fabricated by SEI;
- machine is cooled by low cost, easy-to-handle liquid nitrogen;
- very low noise;
- very low leakage magnetic flux;
- the surface of the machine housing is kept at a room temperature and, therefore, can be installed and used in any place.

The rotating armature is located in the center of the motor between twin stators. The coreless armature winding consists of 6 pancake coils wound with a copper wire. Coreless winding does not produce any cogging torque. The motor is furnished with slip rings and brushes to supply the heavy electric current to each of the armature phase windings, which is the main disadvantage.

The stationary eight-pole field excitation system is placed in fiberglass reinforced plastic (FRP) cryostats. The FRP cryostat is lightweight, strong and does not conduct electric current (no eddy-current losses). The temperature of liquid nitrogen is 66 K in order to increase the field current. There are two field excitation systems (one on each side of the armature) with eight ferromagnetic poles and eight HTS coils each. The pole cores are concentric with respect to the HTS coils, which results in limited penetration of the magnetic flux into the coils. The pole cores protrude outside the cryostat where they

Table 8.5. Specifications of Japanese 12.5 kW HTS disc type motor [187, 190].

Rated output power, kW	12.5
Rated torque, Nm	1194
Short time duty power, kW	62.5
Speed, rpm	100
Armature phase rms current, A	30
Armature current density, A/mm^2	5.0
Liquid nitrogen temperature, K	66
Diameter, m	0.65
Length, m	0.36
Number of turns of HTS field excitation coil	330
Dimensions of HTS BSC wire, mm	4.3 × 0.22
Diameters of HTS coil, mm	ID = 136, OD = 160
Filed excitation pole core diameter, mm	96
Number of turns of armature coil	850
Diameter of armature coil, mm	208
Diameter of armature copper wire, mm	2

are mechanically fixed to the disc-shaped ferromagnetic yokes, which create a back path for the magnetic flux. The cores are wound with a ferromagnetic ribbon to reduce the eddy current losses. Since cores are outside the cryostat, the loss to cool cores is zero.

The maximum magnetic flux density in the core is only 1.8 T, so that the magnetic circuit is unsaturated. The maximum normal component of the magnetic flux density penetrating to HTS coils is 0.29 T, i.e., below the value corresponding to critical current.

The assembled 12.5 kW THS motor and pod propulsor are shown in Fig. 8.20. Performance characteristics obtained from laboratory tests are shown in Fig. 8.21. The maximum efficiency is 97.7% and it remains practically constant for the load from 60 to 210 Nm. However, the cooling loss is not taken into account. The motor was loaded far below the rated torque, which is 1194 Nm (Table 8.5). The maximum shaft torque at laboratory tests was only 210 Nm, which corresponds to 2.2 kW shaft power at 100 rpm. The frontier group justifies it by the limit on the load equipment.

The axial flux HTS motor has a small size and is characterized by low energy consumption. It is expected to find potential applications not only in pod propulsors, but also in railcar propulsion systems, wind generators and large industrial electromechanical drives, e.g., steel rolling mills. The group has a plan to develop a commercial 400 kW, 250 rpm HTS motor for coasting vessels. Long term vision includes development of larger motors 2.5 MW and 12.5 MW. To improve current density, coated YBCO wires will replace BSCCO wires.

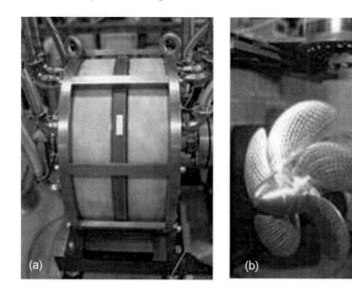

Fig. 8.20. Japanese 12.5 kW HTS motor for pod propulsor: (a) assembled motor; (b) pod propulsor integrated with the motor [187, 190].

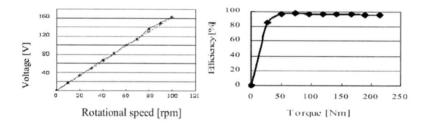

Fig. 8.21. Performance characteristics of the 12.5 kW HTS machine obtained from laboratory tests: (a) voltage versus speed (generating mode); (b) efficiency versus shaft torque (motoring mode) [187, 190].

HTS induction motor

A research group from Kyoto University reported that a small *HTS induction motor* (IM) was successfully run in 2006 [125, 128]. HTS BSSCO wire has been used for cage type rotor winding. Both the rotor bars and end rings have been fabricated using Bi-2223 Ag multifilamentary tapes. Rotor bars and end rings have been designed in a different way (Table 8.6). The narrow width tape, i.e., 2.0 mm, is utilized for rotor bars, because of the restriction of the space in the rotor slot. A bundle of four tapes are connected in parallel by using a solder to make the rotor bar. The total critical current is 100 A (25 A × 4 turns).

Large currents are induced in the end rings as compared with the rotor bars. For this purpose, a tape that has large critical current, ($I_c = 90$ A at 77.3 K and self field) is utilized. The HTS end rings are designed in such a way as to be always in the SC state, i.e., the induced current is less than I_c. The end ring consists of the Bi-2223 Ag tape with 6 turns, i.e., its total critical current is about 540 A (90 A × 6 turns).

Table 8.6. Specifications of Bi-2223 Ag multifilamentary tapes utilized for the squirrel cage rotor windings [125, 128].

Specifications	Rotor bars	End rings
Width, mm	2.0	4.3
Thickness, mm	0.20	0.22
Cross section area, mm^2	0.40	0.95
Critical current I_c at 77.3 K, self field, A	25	90

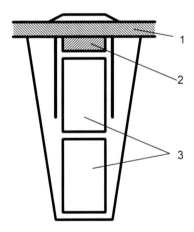

Fig. 8.22. Schematic diagram of a rotor slot with rotor bar and end ring. 1 — HTS end ring (Bi-2223 Ag tape), 2 — HTS rotor bar (Bi-2223 Ag tapes), 3 — insulated copper bars [125, 128].

The HTS cage rotor has a laminated silicon steel core. To minimize the cost of prototyping, the stator of a commercial 3-phase, 4-pole, 1.5 kW IM has been utilized. Thus, the rotor winding is only replaced by the HTS wire and then, the simple stator of a conventional IM is implemented. Fig. 8.22

Fig. 8.23. HTS IM: (a) HTS rotor; (b) assembled machine [125, 128]. Photo courtesy of T. Nakamura, Kyoto University, Japan.

illustrates the schematic diagram of the cross section of the rotor slot. The HTS Bi-2223 rotor bar is put at the outermost surface of the rotor, i.e., near the air gap, in order to reduce the leakage magnetic flux. Insulated copper bars are also inserted in the slot in order to support and cool the rotor bars. Bi-2223 end rings are connected with the rotor bars by using a solder. Fig. 8.23 shows the HTS rotor and assembled HTS motor.

The fabricated IM has been immersed in liquid nitrogen, and then tested at no-load and light load conditions. Before the experiment, the lubricating oil of the roller bearings has been removed by using acetone to avoid freezing. The mechanical losses in the bearings have increased due to the friction. Rotating speed and torque have been measured using a torque–speed meter.

The performance of the IM was also estimated on the basis of the electrical equivalent circuit taking into account the nonlinear E–J curve (electric field versus current density curve) of Bi-2223 Ag tape. It was shown theoretically and experimentally that there is a minimum voltage for starting. The HTS IM also has large starting torque and accelerating torque as compared with the conventional motor (Fig. 8.24). Furthermore, the synchronous torque can be developed by trapping the magnetic flux in the cage winding. This interesting performance is possible only by using HTS IM.

HTS synchronous motor rated at 15 kW

A joint Research and Development Group funded by New Energy and Industrial Technology Development Organization (NEDO) has developed YBCO based 15 kW HTS synchronous motor [88, 89]. This synchronous motor uses stationary HTS field coils and regular copper wire armature winding in the rotor. Fig. 8.25 shows the stationary HTS field excitation system, rotating armature and assembled motor on a test bed. Fig. 8.26 shows racetrack type YBCO field excitation coils.

Specifications of YBCO field excitation winding are listed in Table 8.7. Specification of the 15 kW HTS synchronous motor are given in Table 8.8.

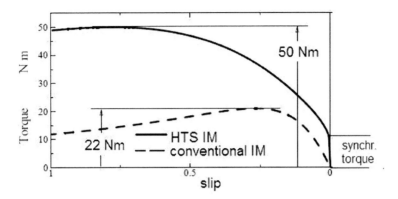

Fig. 8.24. Estimated torque-speed curves of HTS and conventional induction motors [125, 128].

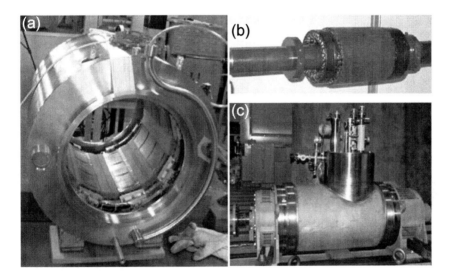

Fig. 8.25. Japanese 15 kW HTS synchronous motor: (a) stationary HTS field excitation system; (b) rotating armature; (c) assembled motor under tests [88, 89]. Photo courtesy of M. Iwakuma, Kyushu University, Fukuoka, Japan.

Fig. 8.26 shows how racetrack type HTS YBCO coils for the stationary field excitation winding have been designed and fabricated.

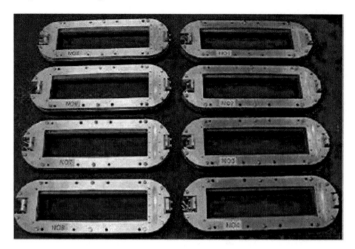

Fig. 8.26. Racetrack type HTS YBCO coils for a stationary field excitation winding of a 15 kW synchronous motor [88, 89]. Photo courtesy of M. Iwakuma, Kyushu University, Fukuoka, Japan.

Table 8.7. Specifications of YBCO field excitation coil of a 15 kW HTS motor [88, 89].

Coil type	Race track type
HTS wire	YBCO 10 mm x 0.1 mm (IBAD-PLD) length 240 m width 1 cm, $I_c > 150$ A at 77K self field
End turn radius, mm	35
Length of straight section, mm	250
Coil width, mm	10.3
Coil thickness, mm	10.5
Rated design current, A	300
Number of turns	35.5
Unit magnetic field at 300 A •maximum flux density in parallel direction, T •maximum flux density in perpendicular direction, T •maximum flux density in the center, T	0.40 0.35 0.11
Inductance, mH	0.47
Wire length (theoretical), m	27

8.1.10 Bulk HTS machines

Another Japanese industry-academia group[2] has developed a 15 kW, 720 rpm, coreless disc type HTS motor, similar to that shown in Figs 8.19 and 8.20. *Kitano Seiki* is marketing this motor [100].

---

[2] *Fuji Electric Systems*, Fukui University, *Hitachi*, Ishikawa-hajima Heavy Industry (IHI), *Kitano Seiki*, *Nakashima Propeller Co., Ltd.*, Niigata Motors, SEI, *Taiyou Nissan* and Tokyo University of Marine Science and Technology.

Table 8.8. Specifications of 15 kW HTS synchronous motor [88, 89].

Type	Rotating armature, stationary HTS field windiing
Number of phases	3
Rated output power, kW	15
Rated speed, rpm	360
Number of poles	8
Voltage, V	360
Frequency, Hz	24
Armature current, A	32
Field winding current, A	280
Armature wire diameter, mm	1.3
Auxiliary equipment	Rotating encoder
Mass, kg	900

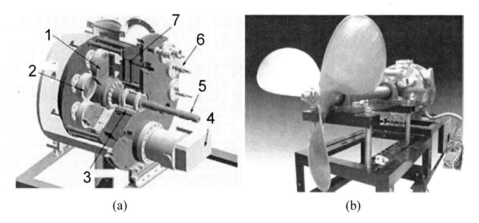

(a) (b)

Fig. 8.27. Disc type brushless machine with bulk HTS rotor marketed by *Kitano Seiki*, Japan: (a) computer 3D image; (b) motor and propeller on test bed. 1 — HTS bulk magnet, 2 — armature coil, 3 — magnetic seal unit, 4 — vacuum pump, 5 — liquid nitrogen inlet, 6 — armature winding terminal, 7 — coil cooling layer.

The axial gap SC motor uses GdBaCuO bulk HTS for rotor excitation system and HTS tapes for the stator (armature) winding. The rotor system is in the central position of the machine and the twin stator (armature) is on both sides of the rotor (Fig. 8.27). The rotor has 8 round poles, 26-mm diameter each. The armature winding has 6 poles. The rotor assembly is 0.3 m long and 0.5 m in diameter. The magnetic flux density in the air gap is 3 T at 77K under field cool magnetization. The bulk HTS poles are magnetized inside the machine using copper coils. The armature system is stationary so that there are no slip rings, only armature leads. The rotor field excitation system with bulk HTS does not require any leads. No electric power supply is needed to maintain the magnetic field. The machine can operate both as a motor and generator. The prototype has been designed for marine pod

Fig. 8.28. *Oswald Elektromotoren GmbH* SRE70-20 bulk HTS machine. Photo courtesy of *Oswald Elektromotoren*, Miltenberg, Germany [159].

propulsors. Potential application to medical, biomedical, and environmental protection are under consideration.

Oswald Elektromotoren, Miltenberg, Germany has reported a construction of 2-pole and 4-pole SC reluctance motors using YBCO bulk material for the field exclusion zone of each pole [159]. Several motors have been built (Fig. 8.28) and operated in liquid nitrogen at different frequencies (speeds) and loads, achieving a mechanical output power of approximately 10 kW. Fig. 8.29 shows power up to 11.3 kW. This corresponds to a maximum force density related to the rotor surface of 40 000 to 69 000 N/m^2, a value which is significantly higher than that for conventional IMs (<20 000 N/cm^2). The small size and low mass of such motors are considered to be important advantages for special applications.

8.1.11 HTS synchronous generator built in Russia

NII Electromach, Saint-Petersburg, Russia, has built a prototype of a high voltage 5 MVA, 40-kV, HTS synchronous generator shown in Figs 8.30 and 8.31 [10, 44]. This generator has been developed for a future operation with a d.c. transmission line via solid state converter.

The machine has an HTS field excitation winding, stator ferromagnetic core (yoke) and slotless stator winding. All fastenings of rotor winding are made of nonmagnetic steel. Slotless high voltage winding is cost effective for high voltage transmission line since the step-up transformer is not necessary [10].

The rotor is a rotating cryostat (liquid nitrogen) carrying an HTS winding consisting of racetrack modules as shown in Figs 8.32 and 8.33. The racetrack winding is made of HTS tape.

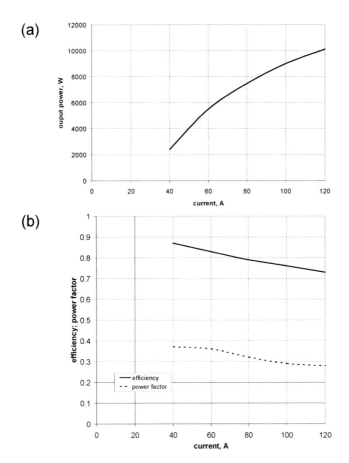

Fig. 8.29. Output power, efficiency and power factor versus current at 1500 rpm for *Oswald Elektromotoren* SRE70-20 bulk HTS motor [159].

The stator (armature) winding consists of multilayer saddle-type concentric coils (Fig. 8.34). The advantages of such a winding include a simple geometry, manufacture and assembly as well as improved cooling conditions. Stator coils have been tested at 100 kV [10].

8.1.12 HTS d.c. homopolar generator

General Atomics, San Diego, CA, U.S.A. claims that an SC *d.c. homopolar motor* is considered to be a conceptually superior alternative to a.c. motors. According to *General Atomics*, a d.c. SC homopolar motor is:

- quieter, smaller, and lighter than a.c. motors;
- more efficient;

Fig. 8.30. High voltage HTS 5 MVA synchronous generator built in Russia under test [10].

- suited to simpler and less costly electrical distribution architectures of naval vessels;
- its control is more straightforward and simpler.

Faraday's disc is the simplest d.c. homopolar machine. A copper or aluminum disc rotates between U-shaped PM poles. One brush collect the current from the shaft and the second brush is placed close to the outer diameter of the disc. PMs can be replaced by d.c. coils. In more practical solutions, the rotor has a shape of a ferromagnetic cylinder coated with a high conductivity layer and magnetized by two coils, as shown in Fig. 8.35. The armature current of a d.c. homopolar generator is inherently high and terminal voltage is inherently low. This is why homopolar generators are used for high current, low voltage plants, e.g., electrolysis of aluminum.

In a d.c. homopolar SC generators the copper field excitation coils are replaced by SC coils immersed in a cryostat. Fig. 8.36 shows a longitudinal section of an HTS d.c. homopolar generator proposed by *General Atomics*.

General Atomics has developed and demonstrated the reliability of conduction cooled SC systems for full-scale homopolar motors for the Navy under high shock and vibration environments. Under contract with the Office of Naval Research (ONR), a team headed by *General Atomics* works on a one-quarter scale prototype of a 3.7 MW d.c. HTS homopolar motor (Fig. 8.37). This project encompasses testing, brush risk reduction, and the design of a higher power-density, ship-relevant, advanced motor.

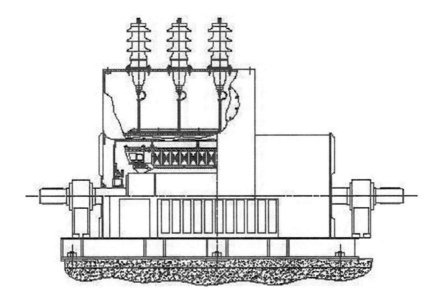

Fig. 8.31. Longitudinal section of high voltage 5 MVA HTS synchronous generator built in Russia [10].

Fig. 8.32. Construction of the racetrack field excitation coil [10].

The main drawback of a d.c. SC homopolar machine is the current collector and brushes, which affect the machine performance and reliability. Dry current collectors and metal fiber brushes that reduce wear rate and maintenance are under development.

The future full scale HTS d.c. homopolar motor rated at 25 MW will have the outer diameter of 2.65 m and length of 3.05 m.

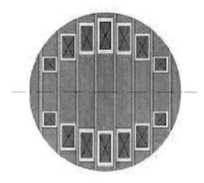

Fig. 8.33. Rotor with field excitation winding [10].

Fig. 8.34. Slotless saddle-type stator winding: (a) two-pole, three-phase winding diagram; (b) saddle-type coil shape; (3) cross section of the three-layer stator winding [10].

8.2 High speed HTS generators

High speed HTS generators are predicted for airborne applications, which include (Section 4.5):

- directed energy weapons (DEWs);
- airborne radars.

High speed HTS generators can also be used as gas turbine driven generators for marine and land applications.

8.2.1 First prototype of high speed superconducting generators

The first prototype of a *high speed SC generator* was with a low temperature superconductor (LTS) rotor. It was developed by *Westinghouse Electric Corporation*, Monroeville, PA, U.S.A., under USAF sponsorship in the early 1970s [28, 145]. The program was directed toward the design, construction and test of a 12 000 rpm, a.c. generator rated at 10 MVA and 5 kV with

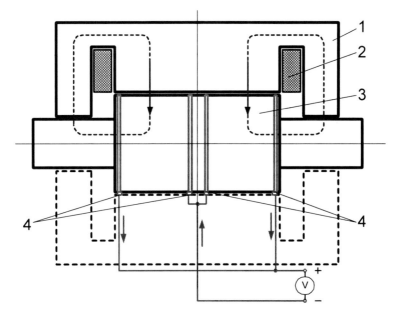

Fig. 8.35. Practical d.c. homopolar generator with cylindrical rotor and two excitation coils. 1 — stator core, 2 — d.c. field excitation coil, 3 — rotor, 4 — brush contact.

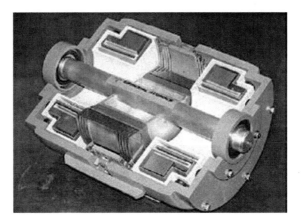

Fig. 8.36. Longitudinal section of HTS d.c. homopolar motor. Photo courtesy of *General Atomics*, San Diego, CA, U.S.A.

LTS field excitation winding. The first phase of this program, which was completed in the early 1974, demonstrated that a four-pole SC rotor could be spun at 12 000 rpm with the field excited to the design current level (Table 8.9). The prototype rotor was cooled down to 4.2 K and cold running tests were conducted up to the overspeed condition of 13 200 rpm.

Table 8.9. Specifications of 10 MVA, 5 kV, 12 000 rpm LTS generator prototype developed by *Westinghouse* [28, 145].

Rating	
Output power	10 MVA
Terminal voltage	5 kV
Voltage regulation	26.5%
Power factor	0.9
Frequency	400 Hz
Number of phases	3 (Wye)
Speed	12 000 rpm
Reactances per unit	
Synchrononous at full load	0.52
Transient	0.44
Subtransient	0.21
Stator a.c. winding	
Current density	25.1 A/mm^2
Stator shield	Hyperco
Dimensions of stator	
Bore seal inner diameter	0.252 m
Inner diameter of conductors	0.263 m
Outer diameter of frame (maximum)	0.483 m
Overall length of stator	0.747 m
Dimensions of rotor	
Rotor outer diameter	0.248 m
Outer diameter of field winding	0.203
Bearing-to-bearing length	0.757 m
Superconductor	
Type	NbTi in copper matrix
Critical current	$I_c = 850$ A at 4 T and 4.2K
Size	0.94 mm × 1.4 mm
Number of filaments	438
Twist pitch	8.4 mm
Filament diameter	0.36 mm
Cu:Sc	2:1
Critical current at 4 T	830 A
Field self inductance	1.22 H
Field current	246 A
Number of field turns per pole	1250
Mass	
Copper in winding	85 kg
Stator shield	134 kg
Total mass of stator	281 kg
Superconductor	34 kg
Total mass of rotor	145 kg
Total mass of generator	426 kg

Fig. 8.37. One-quarter scale HTS 3.7 MW d.c. homopolar motor under construction. Photo courtesy of *General Atomics*, San Diego, CA, U.S.A.

The rotor consists of an internal low temperature structure or core that supports the four SC coils and locates the power lead and helium management system. Surrounding this structure is an electro-thermal shield, which acts both as the eddy current shield and radiation shield.

The four pole structure was made up of four essentially identical wound coils. The SC wire of each coil was coated with a thermoplastic (polybondex) insulation. After completion of the winding fabrication, the coil and winding fixture was heated to a sufficient temperature to soften the thermoplastic and then cooled. The coil, after removal from the winding fixture, was then in a rigid, bonded form, which could be easily handled as a unit [28, 145]. After each coil was completed, it was mounted in a test fixture and tested for maximum current and charging rate capability at 4.2 K with liquid helium under natural convection condition. While the coils differed somewhat in their individual performance, they all easily met the requirement of reaching the rated current.

The field excitation winding was cooled by a liquid helium at 4.2 K. Interconnected longitudinal and radial helium passages have been designed for natural convection cooling.

The stator is of conventional construction with a bore seal, armature winding, insulation, ferromagnetic shield and aluminum frame. The armature winding is assembled by laying twelve identical coils around the bore seal with each coil representing a phase group for one pole [28, 145].

8.2.2 Homopolar generators with stationary superconducting winding

There is a strong interest in a multimegawatt electric power systems (MEPS) with HTS generators for DEWs [21] (Section 4.5). A prototype of the *HTS homopolar generator* is under development at *General Electric* (GE) Global Research Center, Niskayuna, NY, U.S.A. [175].

A homopolar generator with a *stationary HTS field excitation coil*, a solid rotor forging, and stator liquid cooled air-gap armature winding is a promising alternative (Fig. 8.38). In a homopolar machine a d.c. stationary coil embraces and magnetizes a salient pole rotor. The stationary ring-shaped HTS field excitation coil is placed around the steel rotor forging. Salient poles of the twin rotor are shifted by one pole pitch at either end (Fig. 8.39). The stationary field coil magnetizes the rotor salient poles. The rotor magnetic field rotates with the rotor speed, so that the homopolar generator behaves as a synchronous generator.

Fig. 8.38. Stationary HTS winding with solid steel rotor forging of a homopolar generator. Courtesy of GE, Niskayuna, NY, U.S.A. [175].

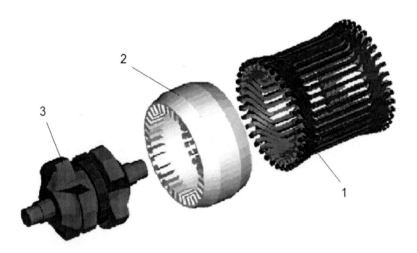

Fig. 8.39. Homopolar generator with stationary HTS field winding. 1 — armature winding, 2 — stator yoke, 3 — rotor with offset poles. Courtesy of GE, Niskayuna, NY, U.S.A. [175].

Construction of a stationary HTS field excitation winding separated from the rotor core has several advantages [175], i.e.,

- no large centrifugal forces acting on the stationary HTS field excitation winding;
- the HTS coil support is simple;
- more reliable stationary HTS coils can be fabricated based on BSCCO or YBCO HTS tapes than in the case of moving HTS coils;
- there are no slip-rings for transferring the excitation current to the HTS field excitation winding;
- there are no rotating brushless exciters and diodes;
- the voltage across the stationary HTS field excitation winding is more predictable, so that it is easier to detect quench and protect the HTS winding using reliable protection devices;
- the thermal insulation between the HTS field excitation winding and ambient is stationary and can be improved because of lack of centrifugal loads and simpler winding support;
- the cryostat for the HTS winding is stationary, so that there is no need for a transfer coupling to introduce a cooling medium into the rotating cooling circuit.

The main drawbacks are:

- high leakage magnetic flux of the stationary d.c. coil;
- larger size of homopolar machine as compared with an equivalent wound rotor synchronous machine.

According to GE researchers, the HTS homopolar generator can offer higher power density for the high speed application than a synchronous generator with HTS rotor [175].

8.2.3 Design of HTS rotors for synchronous generators

There are two types of the *rotor HTS field excitation windings* of synchronous machines: (a) coreless winding and (b) salient pole winding with ferromagnetic core. For high speed generators a *rotor with ferromagnetic core*, similar to that used in low speed HTS machines, seems to be a more reasonable practical solution. A ferromagnetic rotor reduces the magnetizing current and magnetic flux density in HTS wires. There are, in general, many common constructional features of low speed and high speed HTS rotors. The rotor dynamics and protection against centrifugal stresses of field excitation systems is much more serious in the case of high speed HTS rotors. HTS rotor constructions with ferromagnetic cores, similar to those of low speed HTS machines, can be leveraged to high speed HTS machine technology.

Resarchers from the University of Southampton, U.K. propose magnetic flux diverters in the rotor [5, 6]. Separate ferromagnetic rings are placed between the coils. These rings divert the magnetic flux around the coils and in this way reduce the normal component of the magnetic flux density in the superconductor. The FEM analysis shows that the ferromagnetic rings virtually eliminate the normal field component in the coils. Thus, the hybrid design retains the advantage of placing the coils in slots, while allowing the coils to be wound separately and fitted to the core later. This greatly simplifies the construction of the rotor.

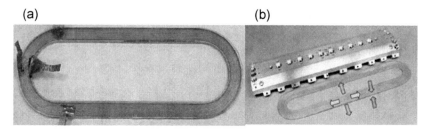

Fig. 8.40. HTS rotor excitation winding of a synchronous machine fabricated by *American Superconductors*: (a) coils; (b) coils and ferromagnetic pole shoe. Photo courtesy of *American Superconductors*, MA, U.S.A.

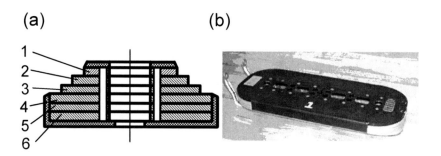

Fig. 8.41. HTS excitation winding of a 150-kW synchronous motor: (a) cross section of a single pole winding (b) single coil. Numbers of turns: coil 1 – 70, coil 2 – 114, coil 3 – 146, coil 4 – 172, coil 5 – 136, coil 6 – 86. Courtesy of *Rockwell Automation*, Greenville, SC, U.S.A.

Fig. 8.40 shows an HTS field excitation winding of a synchronous machine fabricated by *American Superconductors*, Devens, MA, U.S.A.

Fig. 8.41 shows HTS coils of a 150 kW, four-pole synchronous motor built and tested in 1996 by *Rockwell Automation*, Greenville, SC, U.S.A. [167]. Six pancake BSCCO coils connected in series are mounted on each pole. The axial length of each coil is approximately 400 mm. Coils have been stacked on an

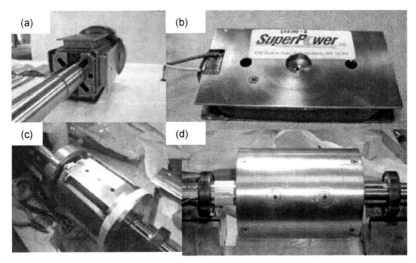

Fig. 8.42. HTS rotor assembly: (a) rotor core without coils; (b) YBCO coils; (c) HTS field coils installed prior to installation of pole faces; (d) completed rotor with shield. Courtesy of *Rockwell Automation*, Greenville, SC, U.S.A.

aluminum mandrel and the whole system has been potted with epoxy to form an integral structure.

The HTS rotor of a smaller, 900 W, 1800 rpm, four pole synchronous machine built and tested by *Rockwell Automation* is shown in Fig. 8.42.

8.3 Market readiness

This section has been prepared on the basis of independent analysis done in 2006 by *Navigant Consulting*, Burlington, MA, U.S.A. [132]. *Navigant Consulting* predicts that the strongest *near term markets* for HTS are not in power utility and energy applications. The following applications can emerge very soon, i.e.,

- military and science applications value the technology attributes most;
- other applications, such as transportation are likely to be important;
- niche applications will likely be key early markets in energy, but more experimentation is required.

Commercial success for HTS technology in the important energy and utility applications will take longer than previously predicted. 2G HTS wire technology does not currently meet any of the application requirements today that will support a commercial market (Table 8.10). However, laboratory research has proven that 2G HTS wire can achieve the required performance targets.

Table 8.10. Wire performance requirements for various utility – industry consensus. Compiled by *Navigant Consulting*, Burlington, MA, [132]. Original data according to [28, 30].

Application	J_c A/cm^2	Flux density T	Temp. K	I_c A	Wire length m	Strain %	Bend radius m	Cost $/kA-m
Large industrial motor (750 kW)	10^5	4 to 5	30 to 77	100 to 500	> 1,000	0.2 to 0.3	0.1	10 to 25
Utility generator	$J_e > 10^4$	2 to 3	50 to 65	125 at operating temp and 3T	> 1,000	0.4 to 0.5	0.1	5 to 10
Synchronous condenser	10^5	2 to 3	30 to 77	100 to 500	> 1,000	0.2	0.1	30 to 70
Transformer	$J_c > 10^6$ $J_e > 12\,500$	0.15	70 to 77	> 100 at 0.15 T	> 1,000	0.3	0.05	10 to 25
Fault current limiter (FCL)	10^4 to 10^5	0.1 to 3	70 to 77	300	> 1,000	0.2	0.1	30 to 70
Power cable (transmission)	$> 10^5$	0.15	67 to 77	200 at 77 K, sf	> 500	0.4	2 (cable)	10 to 50

Table 8.11. Projection of the commercialization timeline for utility/energy HTS applications according to *Navigant Consulting* Burlington, MA, [132].

Application	Initial system prototypes	Demonstration Refined prototypes	Commercial prototypes	Market entry	Market penetration
Industrial motors	2004	2006	2011	2016	2019
Utility generators	2008	2012	2016	2020	2023
Synchronous condensers	2004	2006	2009	2011	2014
Wind generators	2008	2010	2012	2014	2017
Transformers	2003	2008	2011	2014	2017
Fault current limiters (FCL)	2007	2010	2012	2014	2017
Power cables	2006	2008	2011	2014	2017

8.3 Market readiness

Wire performance and price requirements vary by application, and will drive the timing of market entry.

2G HTS wire cost and performance goals required for early commercial energy and utility applications will not be achieved until after 2010 (Table 8.11). The most important near term utility and energy markets appear to be *synchronous condensers* and *fault current limiters* (FCLs). Most possible market scenarios are described below [132]:

- Owing to the relative clarity and strength of their value propositions, the strongest early markets for HTS are likely to be synchronous condensers and FCLs. Mass markets such as cable, transformers and generators that value low impedance and high energy density will emerge much later.
- New applications in power utilities and energy are likely to value the small and light systems, such as off-shore wind turbines. Other new applications may emerge when there is more opportunity to experiment with the technology.
- It is not fully understood how long it will take to develop these markets, but it is likely to take five to ten years of niche applications and experimentation in most segments before broader, mass markets develop.
- It is not clear today if HTS offers a compelling value proposition in many of the important applications that will demand higher volumes of wire and, as a result, more application studies, demonstrations and government support will be required to develop these markets.

9
Naval electric machines

The concept of *electric propulsion* for ships originated over 100 years ago, but lack of the possibility to control large power caused this idea impossible to put in use. Mechanical propulsion systems with diesel engine were more competitive [2]. Electric propulsion of large marine vessels emerged in the 1980s with availability of large power solid state devices and variable speed drives (VSDs).

Replacement of the mechanical propulsion system with the electric propulsion system increases the useable space in a marine vessel, in particular a warship or a cruise liner [99]. The term *electric ship* generally refers to a ship with a full electric propulsion system (mechanical prime mover, electric generator, solid state converter, electric motor, propeller).

9.1 Background

A major benefit of *integrated full electric propulsion* (IFEP) is the layout flexibility offered through removal of the shaft line. There is considerable freedom in location of prime movers, because they are no longer directly coupled to the propulsors allowing for more effective use of the available space (Fig. 9.1). IFEP also offers cost savings.

Electric ships are more than electric-drive systems and include power generation, distribution, and controls. Newer naval ships require significantly larger amount of energy and power (much greater than commercial ships), which is consumed by pulse weaponry, high power microwave, high-power military loads, energy conversion systems and power delivery system (Fig. 9.2). Other loads may include electro-magnetic assistance launch system (EMALS), communication, computer, radar, and sonar systems and hospitality and service loads.

The synchronous generator driven by a gas turbine or diesel engines electrically connected with electric propulsion motor is nowadays used in a large

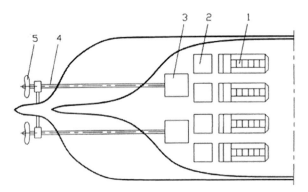

Fig. 9.1. Electric ship propulsion system: 1 — engine-synchronous generator unit, 2 — solid state converter, 3 — large power electric motor, 4 — propeller shaft, 5 — propeller [68].

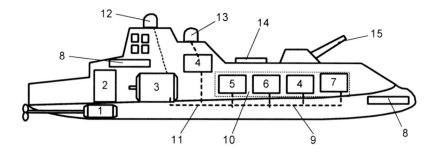

Fig. 9.2. Electric warship major system components according to [147]. 1 — advanced motor and propulsor, 2 — quiet motor drive, 3 — advanced generators, 4 — pulse forming network, 5 — energy storage, 6 — actuators and auxiliaries, 7 — fuel cell stacks, 8 — sensors, 9 — distribution, 10 — integrated power system, 11 — integrated thermal and power management systems, 12 — HPM, 13 — HEL, 14 — electromagnetic vertical launching system, 15 — electromagnetic gun.

variety of vessels. Installed electric propulsion power in merchant marine vessels was in 2002 in the range of 6 to 7 GW. *Azimuthing thrusters* and *podded thrust units* have added superior maneuvering capabilities, dynamic positioning and even some level of intelligence. At present, electric propulsion is applied mainly in the following types of marine vessels:

- cruise vessels;
- ferries;
- dynamic positioning drilling vessels;
- cargo vessels;
- moored floating production facilities equipped with thrusters;
- shuttle tankers;

- cable layers;
- pipe layers;
- icebreakers and other ice–going vessels;
- supply vessels;
- submarines;
- combatant surface ships;
- unmanned underwater vehicles.

The expected compound annual growth rate for electric motors and generators for ship propulsion applications is expected to be more than 20%. Today nearly 100% of all cruise ships and some cargo ships have transitioned to electric motor propulsion systems.

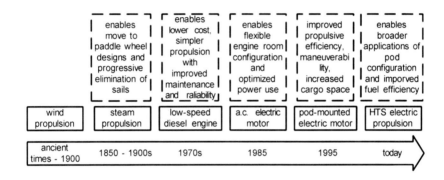

Fig. 9.3. Development of ship propulsion technologies according to *American Superconductor* [9].

Fig. 9.3 shows the history of the development of ship propulsion technologies. The U.S. Navy has proposed electric propulsion system for its new construction ships starting with the DD-21 destroyer. In 1995, *Kaman*, EDC, Hudson, MA, U.S.A. (currently *DRS Technologies*[1], Parsippany, NJ, U.S.A.) developed and tested a prototype of a 2.2 MW, 65-kNm, 1.65 m in diameter and 0.76 m long PM propulsion motor. Key features of *Kaman* design include its removable magnet modules and segmented stators. With these line-replaceable units a motor does not have to be removed from the ship for repair or overhaul. Other marine propulsion programs at *Kaman* have included *Pacific Marine's Midfoil*, i.e., an advanced hull craft that featured twin 750 kW shaft drives, each with a *Kaman* PA57 750 kW PM motor (Table 9.1), a diesel-powered PA57 generator and EI900 inverters [123]. In 2000, the U.S. Navy announced its intention to transition to electric propulsion motors for future navy ships.

---

[1] *Kaman* Electromagnetics Development Center (EDC) sold its electric motor and drive businesses to *DRS Technologies* in 2003.

Table 9.1. Design data of large power, three phase, disc type PMBMs manufactured by *Kaman Aerospace*, EDC, Hudson, MA, U.S.A.

Quantity	PA44-5W-002	PA44-5W-001	PA57-2W-001
Number of poles, $2p$	28	28	36
Number of windings per phase		2	
Output power P_{out}, kW	336	445	746
Peak phase voltage, V	700	530	735
Rated speed, rpm	2860	5200	3600
Maximum speed, rpm	3600	6000	4000
Efficiency			
at rated speed	0.95	0.96	0.96
Torque			
at rated speed, Nm	1120	822	1980
Stall torque, Nm	1627	1288	2712
Continuous current			
(six step waveform), A	370	370	290
Maximum current, A	500	500	365
Peak EMF constant			
per phase, V/rpm	0.24	0.10	0.20
Winding resistance			
per phase at 500 Hz, Ω	0.044	0.022	0.030
Winding inductance			
per phase at 500 Hz, μH	120	60	100
Moment of inertia, kgm^2	0.9	0.9	2.065
Cooling	Water and glycol mixture		
Maximum allowable			
motor temperature, °C		150	
Mass, kg	195	195	340
Power density, kW/kg	1.723	2.282	2.194
Torque density, Nm/kg	5.743	4.215	5.823
Diameter of frame, m	0.648	0.648	0.787
Length of frame, m	0.224	0.224	0.259
Application	Traction, Drilling industry		General purpose

The largest marine electric motors are installed on cruise vessel (formerly translatlantic passenger liner) *Queen Elizabeth 2*. Two 44 MW, 144 rpm, 60 Hz salient pole GEC synchronous motors drive two propeller shafts. Synchronous motors are 9 m in diameter and weigh over 400 t each. The total electric power of *Queen Elizabeth 2* is 95 MW and is generated by nine three-phase 10.5 MW, 10 kV, 60 Hz GEC water cooled, salient pole synchronous generators driven by nine diesel engines.

Queen Mary 2 is the largest, longest, tallest, widest and most expensive passenger cruise vessel ever built (Fig. 9.4). Its power plant includes two gas

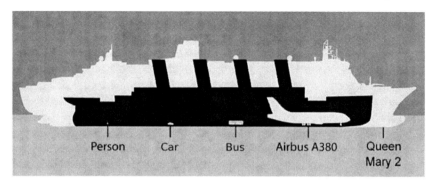

Fig. 9.4. Comparison of *Queen Mary 2* with *Titanic*, Airbus A-380, person, car and bus.

turbines and four diesel engines that produce 118 MW of electricity, enough to power a city of 300 000 people. More than two-thirds of this energy is used to power the propulsion system, as each of four electric motors draws 21.5 MW during full power. *Queen Mary 2* is outfitted with four *Rolls Royce Mermaid*^{TM} pod propulsors, two fixed and two azimuthing, i.e., rotating 360^0.

9.2 Power train of electric ships

The main components of power train of electric ships are prime mover, generator, switchboard, transformer and propulsion motor [2, 35, 38, 99, 107, 162, 181].

(a) *Prime mover*. In most electric ships generator sets are driven by a combustion engine, which is fueled with diesel or heavy fuel oil. Quite often, gas engines, gas turbines, steam turbines or combined cycle turbines are used. Gas turbines are used for higher power ranges, in light high-speed vessels, or where gas is a cost effective alternative, e.g., waste product in oil production, boil-off in liquefied natural gas (LNG) carriers, etc.

(b) *Generator*. Most commercial vessels have a.c. power generation plants with 3-phase synchronous generators. For reduced maintenance nearly all generators are equipped with brushless exciters. The excitation is controlled by an automatic voltage regulator (AVR) that senses the terminal voltage of the generator and compares it with a reference value. Generators are sometimes designed for two-directional power flow, which means that the generator can run as a motor, i.e., the so–called *power take-in - power take out* concept (PTI – PTO). Since the discovery of HTS in 1986, U.S. Office of Naval Research (ONR) and Department of Energy (DoE) have advocated and founded research in HTS synchronous and homopolar generators for on-board electricity generation.

(c) *Switchboard*. Main (generator) switchboards are usually distributed or split in two, three, or four sections, in order to meet redundancy requirements. According to rules and regulations for electric propulsion systems, consequences of one section failing, e.g., due to a short circuit, shall be tolerated. For the strictest redundancy requirements, failures due to fire or flooding, meaning that water and fireproof dividers must be used to segregate each section, shall also be tolerated.

(d) *Transformer*. The purpose of the transformer is to isolate the different parts of the electric power distribution system and divide them into several partitions, normally, to obtain different voltage levels and sometimes also phase shifts. Phase shifting transformers can be used to feed frequency converters, e.g., for variable speed drives, to minimize distorted currents in the electric power grid by canceling the most dominant harmonic currents. Transformers also damp a high frequency conductor emitted noise, especially, if the transformer is equipped with a grounded copper shield between the primary and secondary winding.

(e) *Propulsion motor*. For electric marine propulsion systems IMs, wound rotor synchronous motors, PMBMs of cylindrical construction, axial flux disc type PMBMs, PM transverse flux motors (TFMs) and synchronous motors with HTS field excitation winding can be used.

9.3 Propulsion units

9.3.1 Shaft propulsion

In a *shaft propulsion* system, a variable speed electric motor is connected directly or through a step down gear box to the *shaft of propeller*. A gear coupling allows for using a smaller motor rated at higher speed. The disadvantage of the geared shaft propulsion is the increased mechanical complexity, power losses in a gear train and need for maintenance (lubricating oil). The motor is fed from electric generator, which in turn is driven by a diesel engine or turbine engine (Fig. 9.1).

In diesel–electric vessels, shaft propulsion is typically used where the transverse thrust (at maneuvering) is not needed, or, it can be produced in more cost effective way by tunnel thrusters, or where the propulsion power is higher than that available for azimuthing thrusters. Typically, shuttle tankers, research vessels, larger anchor handler vessels, cable layers, etc., are equipped with propeller shafts.

The shaft line propulsion is always combined with rudders, one rudder per propeller. The propeller is normally a speed controlled fixed pitch propeller (FPP) type. In some applications, the propeller can be designed as a controllable pitch propeller (CPP) type, even if variable speed propulsion motor is used.

9.3.2 Azimuth thrusters

In a standard *azimuth thruster* (Fig. 9.5a) the propeller is rotated $360°$ around the vertical axis, providing multi-directional thrust. The electric motor is installed above the water line and drives the propeller with the aid of a gear transmission system. The thrust is controlled either by constant speed and CPP, variable speed and FPP, or sometimes using a combination of speed and pitch control. Variable speed FPP thruster has a significantly simpler underwater mechanical construction with reduced low-thrust losses as compared to constant speed, CPP propellers. When in-board height of the thruster room is limited, the electric motor is horizontally installed and the azimuth thruster consists of a Z-type gear transmission. When the height of the thruster room allows, vertically mounted motors and L-shaped gear transmissions are simpler and more energy efficient (lower power transmission losses). Standard azimuth thrusters are usually rated up to 7 MW.

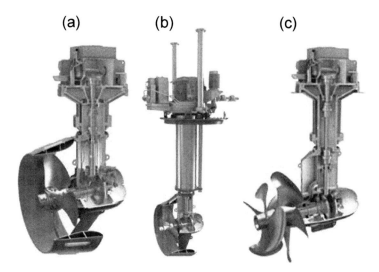

Fig. 9.5. *Ulstein Aquamaster* azimuthing thrusters manufactured by *Rolls-Royce*: (a) standard thruster, (b) retractable thruster, (c) azimuthing thruster with counter-rotating propeller [162]. Courtesy of *Rolls-Royce Marine AS*, Ulsteinvik, Norway.

The so–called *retractable thruster* (Fig. 9.5b) provides fast hydraulic lifting and lowering of the unit. Retractable thrusters have the same main components as standard azimuth thrusters.

Some manufacturers provide azimuth thrusters with *dual propellers*, either on the same shaft, or with counter-rotating propellers (Fig. 9.5c). A *counter-rotating propeller* utilizes the rotational energy of the jet stream produced by one propeller to create the thrust of the other propeller that rotates in the opposite direction. Thus, the hydrodynamic efficiency increases.

9.3.3 Pod propulsors

Similar to the azimuth thruster, the *pod propulsor* can freely rotate and produce thrust in any direction (Fig. 9.6). The electrical power is transferred to the motor via flexible cabling or slip rings to allow a $360°$ operation. Unlike the azimuth thruster, the pod propulsor has the electric motor submerged under the vessel hull and directly integrated with the propeller shaft inside a sealed pod unit. The transmission efficiency is higher than that of an azimuth thruster because of lack of mechanical gears.

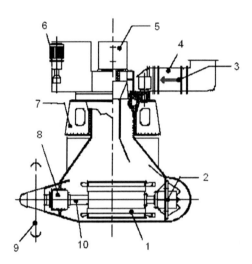

Fig. 9.6. Pod propulsor. 1 — electric motor, 2 — bearing, 3 — air cooling, 4 — ventilation unit, 5 — slip ring unit (power and data transmission), 6 — hydraulic steering unit, 7 — installation block, 8 — bearing, shaft seal, 9 — FPP, 10 — shaft line [2]. Courtesy of *ABB AS Marine*, Oslo, Norway.

The marine pod can be designed for pushing or pulling operation. Especially, the pulling type increases the hydrodynamic efficiency of the propeller and reduces the risk for *cavitation*[2], which means reduced noise and vibration. A podded unit can rotate in both forward and aft directions, if the thrust bearings allows for it.

Pod propulsors are available in power ranges from approximately 1 MW up to 30 MW and have been used for more than a decade in cruise vessels, icebreakers, service vessels, tankers and semi-submersible drilling units. The largest pod propulsors are the *Azipod®* system from *ABB Oy*, Helsinki, Finland and the *Mermaid*™ system from *Rolls Royce* owned *Kamewa®* and

---

[2] Cavitation is the formation of partial vacuums in a liquid by a swiftly moving solid body, e.g., a propeller. It limits the maximum speed of propeller-driven ships to between 55 and 65 km/h (30 and 35 knots).

9.3 Propulsion units   221

Fig. 9.7. Four Rolls-Royce *Mermaid*™ pod propulsors of *Queen Mary 2*.

Alstom Powers Motors. Both have modernized the cruise industry and have potential to enter other markets. The *Queen Mary 2* cruise ship is outfitted with four 21.5 MW *Rolls-Royce Mermaid*™ pod propulsors: two are fixed and two can rotate $360°$ (Fig. 9.7). *Siemens-Schottel SSP*, Spay/Rhein, Germany, CPP type pod propulsors are in range from 1 to 30 MW with propeller sizes varying between approximately 1.8 to 8.0 m.

The *Liberty of the Seas* cruise ship built in 2006 at *Aker Yards*, Finland and owned by *Royal Caribbean Int.*, is propelled by $3 \times 14 = 42$ MW ABB *Azipod*® thrusters and $4 \times 3.4 = 13.6$ MW thruster motors. Six 17.6 MVA generators provide 105.6 MVA power. This 160 thousand ton cruise ship is 339 m long, 56 m wide and can carry 3634 passengers and 1360 crew members on 18 decks.

9.3.4 Integrated motor-propeller

The *integrated motor-propeller* (IMP) sometimes called *rim driven thruster* (RDT) consists of a shrouded (hidden) propeller around which the rotor of an electric motor is mounted (Fig. 9.8). The rotor core is hermetically sealed either by canning or encasing as a monolithic structure in polymer. The stator is enclosed in the stator housing and hermetically sealed by a welded can or sometimes embedded in polymer. A submerged water-cooled IMP offers several advantages over ordinary propulsors or thrusters, i.e.,

- the dynamic shaft seal can be eliminated — only static seals are required for power and instrumentation cables;
- the motor thrust bearing do not have to withstand the depth pressure and full propulsion thrust load;

- the motor effectively utilizes passive sea-water cooling, so that an active cooling and heat exchanger system is not necessary;
- sea water is used for bearing lubrication, so that oil lubrication system is eliminated;
- some IMPs, e.g., developed by *Brunvoll* [35] have no central shafts and no supporting struts, so that water inflow to the propeller is more uniform and undisturbed, which is beneficial with regard to efficiency and propeller induced noise and vibration;
- reliability is increased;
- maintenance costs are reduced.

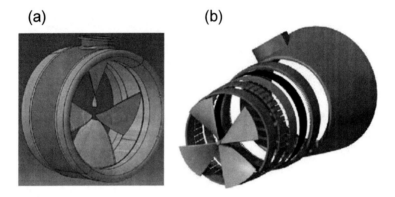

Fig. 9.8. Integrated motor propeller (IMP): (a) outline [35]; (b) exploded view [106].

IMP's use variable speed PMBMs and FPPs. The rotor field excitation system with surface configuration of sintered NdFeB PMs is the most viable option.

IMPs may be applied for all thruster applications with variable speed drives as [35]:

- main propulsion azimuth thruster or auxiliary propulsion thruster;
- retractable azimuth thruster;
- combined tunnel and azimuth thruster;
- tunnel thruster.

IMPs were first introduced by U.S. Navy for submarines, e.g., *Jimmy Carter* SSN23 as submarine auxiliary maneuvering devices [64].

9.4 Generators for naval applications

Electric power for marine electric power train is generated by conventional synchronous generators with electromagnetic excitation. In the future, synchronous generators with HTS excitation system may be used (Chapter 8). HTS generators are expected to be about half (50%) the size and weight of classical synchronous generators.

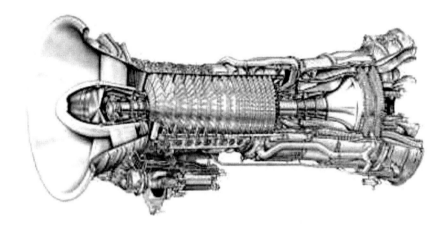

Fig. 9.9. 22 MW, 3600 rpm, 16-stage GE LM2500 gas turbine.

Fig. 9.10. Nine-cylinder, 16.8 MW, 514-rpm, 16V46C-CR Wärtsilä EnviroEngine. Photo courtesy of *Wärtsilä* Corporation, Helsinki, Finland.

224   9 Naval electric machines

Synchronous generators for ship power generation are usually three-phase generators with wound rotor of cylindrical type, similar to turboalternators for large electric power plants. Almost exclusively, brushless exciters are used. The output power is from hundreds of kWs up to over 20 MWs. For example, the biggest passenger liner *Queen Mary 2* is equiped with a 118 MW electric power plant consisting of two GE LM2500 gas turbines, 22 to 25.4 MW each (Fig. 9.9) and four 16V46C-CR Wärtsilä diesel engines, 16.8 MW each at 514 rpm (Fig. 9.10).

9.5 Electric motors for naval applications

9.5.1 Large induction motors

Under contract with *Northrop Grumman* Ship Systems, *Alstom* (formerly GEC-Alsthom) Power Conversion has developed an integrated power system (IPS) consisting of a generating plant, advanced 19-MW IPS IM and motor controller VDM 25000. The U.S. Navy has already tested the 19-MW advanced IM, which will be the baseline for multi-mission surface combatant ship DDG 1000 (also known as Zumwalt-class destroyer).

Fig. 9.11. Large 19 MW, 150 rpm advanced IM. Photo courtesy of *Alstom*, Levallois-Perret, France.

The 19 MW IPS advanced IM has 15-phases and at 15 Hz input frequency develops speed of 150 rpm (Fig. 9.11). Dimensions are 4.8-m length, by 4.5-m width and 4.0 m height.

9.5.2 Large wound rotor synchronous motors

Wound rotor synchronous motors for ship propulsion have salient pole rotors. The largest wound rotor synchronous motor for ship propulsion is rated at 44 MW, 144 rpm, 60 Hz and has 9-m in diameter and weighs over 400 t (Fig. 9.12). Two such motors are aboard the *Queen Elizabeth 2*. The driving power of the synchronous motors is transmitted via a twin shaft arrangement to two five-bladed controllable pitch propellers (CPPs) of 5.8 meter diameter each (Fig. 9.13). The speed is determined by the pitch of the propeller blades. The variable-pitch propellers operate at 144 rpm (cruising speed) or 72 rpm.

Fig. 9.12. One of the GEC 44 MW synchronous motors being lowered through the funnel hatch during 1986–87 major refit of *Queen Elizabeth 2* at the Lloyd Werft yard at Bremerhaven in Germany transforming from a steamship to modern diesel-electric ship.

Comparison of 44-MW *Queen Elizabeth 2* synchronous motor and 19-MW IPS *Alstom* advanced IM is given in Table 9.2.

9.5.3 Large PM motors

First prototypes of rare-earth PM motors rated at more than 1 MW for ship propulsion were built in the early eighties. A review of constructions and associated power electronics converters for large PM motors designed in Germany are described in the author's book [68].

A 36-MW (2 x 18 MW) PMBM (Fig. 9.14), developed by *DRS Technologies* has been tested in 2007/2008 at Land-Based Test Site (LBTS) at the Ships

Fig. 9.13. Controllable pitch propellers (CPPs) of *Queen Elizabeth 2*.

Table 9.2. Comparison of *Queen Elizabeth 2* synchronous motor with IPS *Alstom* advanced IM.

Large a.c. motor for ship propulsion	Power MW	Mass kg	Torque 10^6 Nm	Power-to weight kW/kg	Torque density Nm/kg	Power-to volume MW/m^3
QE2 synchronous motor	44	285 000	2.915	0.15	10.17	0.1
IPS *Alstom* IM for U.S. Navy	19	121 000	1.207	0.16	10.03	0.2

Systems Engineering Station, Philadelphia, PA, U.S.A. Both the motor and solid state converter have modular construction. The PM motor was originally envisioned for the next generation U.S. navy DD(X) destroyer program (now multi-mission surface combatant ship DDG 1000), but technical issues caused delays. The U.S. Navy remains very interested in using PM motor technology in the nearest future.

9.5.4 Axial flux disc type PM brushless motors

Stators of large axial flux PM brushless motors with disc type rotors usually have three basic parts [38]:

- aluminum cold plate;
- bolted ferromagnetic core;
- polyphase winding.

The cold plate is a part of the frame and transfers heat from the stator to the heat exchange surface. The slots are machined into a laminated core wound

Fig. 9.14. Large 36.5 MW PMBM for ship propuslsion. Photo courtesy of *DRS Technologies*, Parsippany, NJ, U.S.A.

in a continuous spiral in the circumferential direction. The copper winding, frequently a Litz wire, is placed in slots and then impregnated with a potting compound.

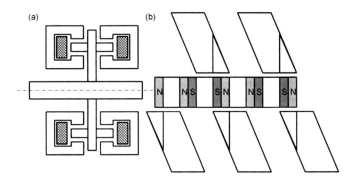

Fig. 9.15. Layout of *Rolls-Royce* TFM: (a) topology; (b) magnetic circuit spread flat.

Fig. 9.16. Prototype of 2 MW *Rolls-Royce* TFM. Photo courtesy of *Rolls-Royce*, Derby, U.K.

Fig. 9.17. Completed stator assembly of 2 MW TFM. Photo courtesy of *Rolls-Royce*, Derby, U.K.

9.5.5 Transverse flux motors

Transverse flux motors (TFMs) have been described in Section 3.1. *Rolls-Royce*, Derby, U.K has invested in developing a high power density electric TFM to allow the application of IFEP technology to the smaller frigate/corvette classes combatant surface ships. Another potential applications of TFMs are nuclear/electric submarine propulsion systems.

Rolls-Royce has selected a double-sided TFM topology shown in Fig. 9.15a. The rotor comprises alternate soft steel laminated pole pieces and PMs (Fig. 9.15b). The magnetization of PMs is in the circumferential direction and aligned in such a way that the pole pieces create alternate N and S poles [85].

Rolls-Royce has designed, fabricated and tested a 2 MW, 2-disc, 2-phase, 76-pole, 195 Hz, 308 rpm demonstrator TFM (Fig. 9.16). The mass of the machine is 13 t and dimensions $1.475 \times 1.5 \times 1.55$ m. The rotor pole stacks are made from anisotropic laminations. Each PM pole is made from 52 pieces of sintered NdFeB bonded together and machined to the final dimensions. The subdivision of PMs into smaller pieces is required to reduce the eddy current losses due to variation of the magnetic flux density with the rotor angle and armature current [85]. The stator cores are bonded into the machined stator frame. The stator coils are arranged in three double layers with inter layer insulation and two connections per layer. Each double layer consists of 14 stranded turns. The stator coil is fitted within the ring of the stator cores and held in coil chairs that are positioned between each stator core (Fig. 9.17). Several improvements to the stator design have been made after laboratory tests.

9.5.6 IMP motors

Induction, wound synchronous and PMBMs are the candidates for IMPs. To maximize the efficiency, increase the torque density and increase the air gap, PMBMs are preferred.

The stator core is radially thin and has large number of slots (Fig. 9.18a). Full ring laminations are mostly used. Non–pressure compensated stator enclosures use structural backup rings to prevent the can collapse [64]. The PM rotor may use surface or embedded PMs (Fig. 9.18b).

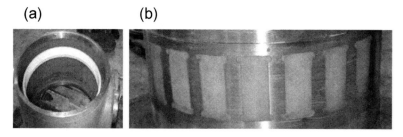

Fig. 9.18. *Brunvoll* PM brushless motor for IMP: (a) stator; (b) rotor. Photo courtesy of *Brunvoll AS*, Molde, Norway [35].

230   9 Naval electric machines

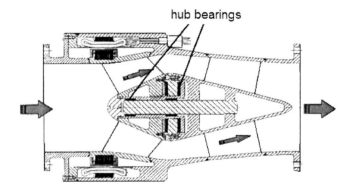

Fig. 9.19. IMP with hub bearings [64].

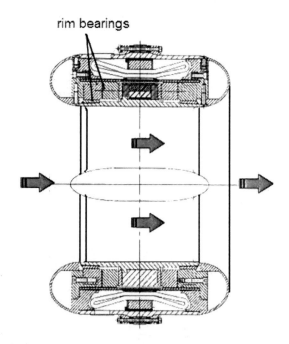

Fig. 9.20. IMP with rim bearings [64].

The rotating propeller/rotor assembly requires bearings to carry loads in the radial and axial directions. The bearing can either be located in the hub (Fig. 9.19) or rim (Fig. 9.20) [64]. Appropriate bearing design is dictated by the application. As a rule, water lubricated hydrodynamic bearings or rolling element type bearings are used. Experiments with magnetic bearings have also been performed (Fig. 9.18a) [35].

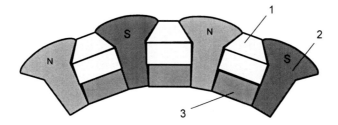

Fig. 9.21. *Curtiss-Wright EMD* large diameter rotor geometry: 1 — PM, 2 — laminated ferromagnetic pole, 3 — nonmagnetic steel. U.S. Patent 6879075.

There were several attempts to find optimum configuration of PM rotor of a large diameter IMP. According to [15], surface mounted PMs and breadloaf PMs with laminated pole shoes exhibited the highest torque per magnet volume. In the case of radially embedded PMs in a laminated core with flux barriers and tangentially inserted PMs in a laminated core the leakage flux limits the torque production [15].

Curtiss-Wright EMD, Cheswick, PA, U.S.A. proposes trapezoidal PMs placed between salient poles (Fig. 9.21). PMs are magnetized tangentially. The rotor is comprised of a nonmagnetic steel ring, laminated ferromagnetic poles between PMs and a corrosion resistant embedding material or a can to seal the rotor from the sea water (U.S. Patent 6879075). The nonmagnetic steel ring is shrunk fit to the propeller shroud. The rotor is totally protected from corrosion by a hermetically sealed boundary with no gaskets or rings [64].

According to *SatCon*, MA, U.S.A. the power density of about 1000 W/kg is achievable [138]. *SatCon* have supplied motor components for IMP (RDT) development programs at the Naval Underwater Warfare Center (NUWC). A two stage potting technique used for a 7.5 kW motor for unmanned undersea vehicle (UUV) provides a good thermal path from the motor windings to the seawater (Fig. 9.22a). *SatCon* has also developed a 300-mm, high-speed, 110 kW PMBM (Fig. 9.22b) for NUWC *Elite* torpedo[3] project. The motor stator is potted directly into the torpedo after-body cone structure. This allows potting of the high current motor leads into the support fins for the propulsor shroud eliminating interconnections and pressure feed-troughs, and the leads are effectively cooled.

In the next project, funded by ONR, *SatCon* selected a 37 kW IMP motor for analysis and demonstration of its operation in large UUV. The objective was to improve on overall power density of the complete propulsor including its hydrodynamic and structural components, bearings, and drive electronics as well as the motor. Computer created images of IMP with 37-kW, 1000-rpm

---

[3] Torpedo is self-propelled guided projectile that operates underwater and is designed to detonate on contact or in proximity to a target.

232   9 Naval electric machines

Fig. 9.22. IMP motors: (a) 7.5 kW, (b) 110 kW [138]. Photo courtesy of *SatCon*, MA, U.S.A.

PMBM are shown in Figs 9.23 and 9.24. The nominal parameters are: IMP diameter 0.53 m, speed 1000 rpm (speed range 500 to 2000 rpm), ambient water temperature 25°C, thrust 3.96 kN, speed of vehicle 22 km/h, hydraulic efficiency 65%, supply voltage 600 V d.c. The motor design is optimized for power density rather than efficiency, which results in reduced weight of the motor itself but also allows minimization of the propulsor duct volume further reducing propulsor weight.

It has been found that arrangement of PMs into Halbach array (4 magnets per pole at 0, 90°, 180°, 270°) can reduce the weight of rotor electromagnetic

Fig. 9.23. Preliminary mechanical model of a 37 kW, 986 W/kg power density, 1000 rpm IMP [138]. Courtesy of *SatCon*, MA, U.S.A.

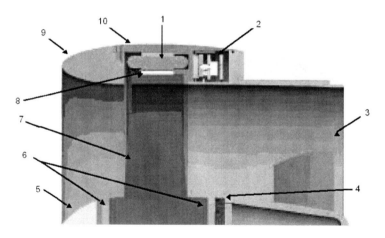

Fig. 9.24. Longitudinal section of a 37 kW, 986 W/kg, 1000 rpm IMP. 1 — stator core and winding of motor, 2 — electronics, 3 — stator, 4 — stub shaft, 5 — hub nose, 6 — bearings, 7 — impeller, 8 — PMs, 9 — nose, 10 — frame [138]. Courtesy of *SatCon*, MA, U.S.A.

components by 40%, allows a radially thinner rim on the rotor assembly, enables a thinner duct (or shroud) and provides lower EMF distortion.

The longitudinal section of the IMP is shown in Fig. 9.24. The motor stator and drive electronics are mounted in the outer duct of the propulsor in a compact and very efficiently cooled unit. The mechanical structure of the rotor consists of an aluminum frame with syntactic foam to create the hydrodynamic shapes. A simple, single bridge, 3-phase solid state converter architecture was selected for the preliminary design in order to maximize power density. Utilizing very compact *SatCon*–developed IGBT packages (1200 V IGBT and anti-parallel diodes rated at 75 A continuous current with a 70^0C heat sink) allow a six switch, three-phase bridge to be accommodated in the propulsor duct.

9.5.7 Superconducting motors

Superconducting synchronous motors are smaller and more efficient than classical synchronous motors for ship propulsion. HTS motors have been described in Chapter 8.

10

Scenario for nearest future

It is expected that the development of electric machines and associated power electronics in the next few years will be stimulated by the following large scale applications:

- computer hardware;
- residential and public applications;
- land (HEVs and EVs), sea and air transportation;
- renewable energy generation.

Further development of electric machines is not limited to these four major areas of applications. For example, artificial electromechanical organs, especially artificial hearts, ventricular assist devices, surgical robots, artificial limbs and other clinical engineering apparatus require very reliable, lightweight, super efficient motors and actuators operating at temperature that never exceeds the temperature of human blood (36.8^0C ± 0.7 or 98.2^0F ± 1.3).

10.1 Computer hardware

10.1.1 Hard disc drive motors

PMBMs in computer hardware are used as *disc drive motors* and *cooling fan motors*. The data storage capacity of a hard disc drive (HDD) is determined by the *aerial recording density* and *number of discs*. The aerial density of HDD has increased from 6 Gbit/cm^2 = 38.7 Gbit/in^2 in 1999 to 20.5 Gbit/cm^2 = 132 Gbit/in^2 in 2006 and 28 Gbit/cm^2 = 180 Gbit/in^2 in 2007.

Mass of the rotor, moment of inertia and vibration increase with the number of discs. Circumferential vibrations of mode $m = 0$ and $m = 1$ cause deviations of the rotor from the geometric axis of rotation [92, 117]. Drives with large number of discs have the upper end of the spindle fixed with a screw to the top cover (Fig. 10.1a). This *tied construction* reduces vibration

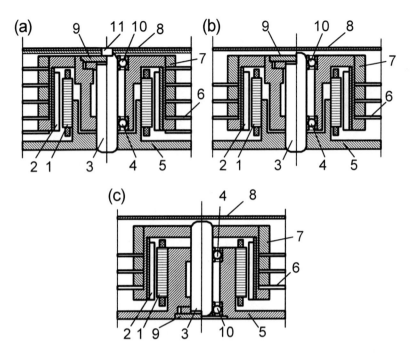

Fig. 10.1. Construction of spindle motors for HDDs: (a) tied type, (b) untied type with fixed shaft, (c) untied type with rotary shaft. 1 — stator, 2 — PM, 3 — shaft, 4 — ball bearing, 5 — base plate, 6 — disc, 7 — disc clamp, 8 — top cover, 9 — thrust bearing, 10 — radial bearing, 11 — screw.

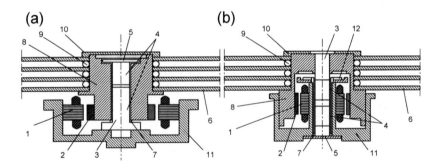

Fig. 10.2. Construction of fluid dynamic bearing (FDB) spindle motors for HDDs [8]: (a) fixed-shaft spindle motor, (b) rotating-shaft spindle motor. 1 — stator, 2 — PM, 3 — shaft, 4 — radial bearing, 5 — thrust bearing, 6 — disc, 7 — stopper/seal, 8 — hub, 9 — spacer, 10 — clamp, 11 — base plate, 12 — attractive magnet.

and deviations of the rotor from the centre axis of rotation. For smaller number of discs the so–called *untied construction* with fixed shaft (Fig. 10.1b) or rotary shaft (Fig. 10.1c) has been adopted.

Special design features of spindle motors are their high starting torque, limited current supply, reduced vibration and noise, physical constraints on volume and shape, contamination and scaling problems. The acoustic noise is usually below 30 dB(A) and projected mean time between failure (MTBF) is 100 000 h.

Drawbacks of ball bearings include noise, low damping, limited bearing life and non repeatable run out [117]. The HDD spindle motor has recently been changed from ball bearing to *fluid dynamic bearing* (FDB) motor. Contact-free FDBs are cogging torque free, produce less noise and are serviceable for an extended period of time (Fig. 10.2).

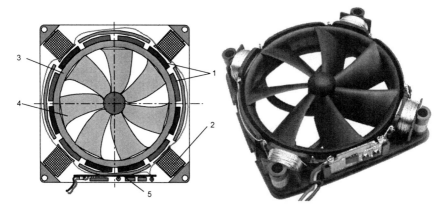

Fig. 10.3. Tip driven fan (TDF): 1 — surface PMs, 2 — stator salient pole, 3 — rotor rim, 4 — fan blades, 5 — control electronics. Photo courtesy of *Yen Sun Technology Corporation*, Kaohsiung, Taiwan.

10.1.2 Cooling fan motors

Computer cooling fans are driven by simple PMBMs that have external rotor ring magnet and internal salient pole stator. The rotor is integrated with fan blades.

The latest technology for cooling fans, the *tip driven fan* (TDF) uses rim mounted electromagnets (salient poles) to spin the fan blades (Fig. 10.3). The hub is free of any driving components, allowing to be sized down to the bearings alone. The central hub area of a TDF is reduced by at least 75% and airflow is increased by 30%. This is important in applications such as CPU heat sinks, which require as much airflow directed down as possible. Noise is also reduced because this design eliminates the tips of the fan, which are the main source of aerodynamic noise in a traditional cooling fan.

Table 10.1. PM vibration motors manufactured by *Precision Microdrives*, London, U.K.

Parameter	Cylindrical type			Disc type	
Model	304-001	304-002	304-101	312-101	312-103
Volts, V	1.3	3	3	3	3
Diameter, mm	4	4	4	12	12
Length, mm	11	8	11	3.4	2.7
Weight, g	1	1	1	5	5
Speed, rpm	10 000	11 000	9000	9000	9000
Current, mA	85	100	60	60	90
Starting voltage, V	1.0	2.4	2.0	2.0	2.4
Starting current, mA	125	120	70	85	120
Terminal resistance, Ω	10.4	–	43	35	25
Operating range, V	1.0–2.0	2.4–3.6	2.0–3.6	2.0–3.6	2.4–3.6

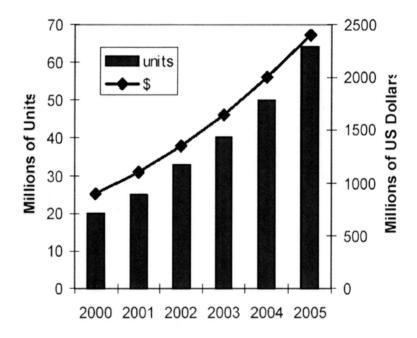

Fig. 10.4. Worldwide market between 2000 and 2005 for motor drives in household applications.

10.2 Residential and public applications

10.2.1 Residential applications

The quantity of small electric motors found in a normal household easily exceeds 50, not including auxiliary electric motors used in gasoline-powered

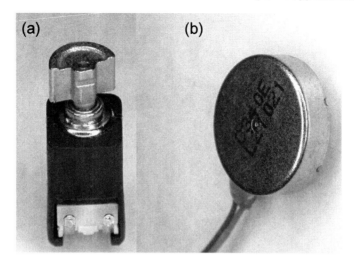

Fig. 10.5. Brush type PM vibration motors: (a) cylindrical (bar) type; (b) disc (coin) type. Photo courtesy of *Precision Microdrives*, London, U.K.

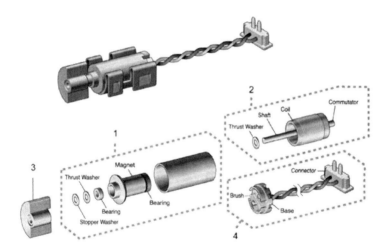

Fig. 10.6. Brush type PM vibration motor of cylindrical construction: 1 — stator assembly, 2 — rotor assembly, 3 — counterweight, 4 — brush assembly. Courtesy of *Samsung Electromechanics*, Suwon City, Gyeonggi, South Korea.

automobiles. These motors should be cost effective, energy efficient, reliable and recyclable (in the future). The role of PMBMs is increasing with the drop in prices of PMs and integrated circuits. SRMs find, so far, a few applications, such as some types of washing machines and air conditioner compressors. Small electric motors are used in the following residential applications:

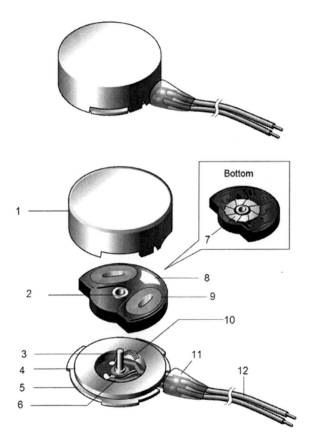

Fig. 10.7. Brush type PM vibration motor of pancake construction. 1 — enclosure, 2 — bearing, 3 — shaft, 4 — PM, 5 — bracket, 6 — flexible printed circuit (FPC), 7 — commutator, 8 — counterweight, 9 — coil assembly, 10 — brush, 11 — ultraviolet curable adhesive, 12 — lead wire. Courtesy of *Samsung Electromechanics*, Suwon City, Gyeonggi, South Korea.

(a) kitchen equipment,
(b) timepieces,
(c) bathroom equipment,
(d) washers and dryers,
(e) vacuum cleaners,
(f) furnaces, heaters, fans, airconditioners, humidifiers and dehumidifiers,
(g) lawn mowers,
(h) pumps (wells, swimming pools, jacuzzi/whirlpool tubs),
(i) toys,
(j) vision and sound equipment, cameras, mobile phones (vibration motors),
(k) computers
(l) power tools,
(m) security systems (automatic garage doors, automatic gates).

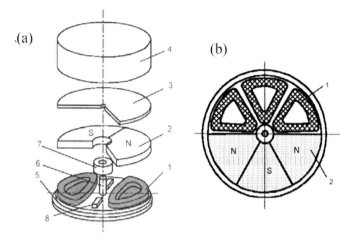

Fig. 10.8. Disc type PM brushless vibration motor for mobile phones [45]: (a) two coil motor; (b) multicoil motor. 1 — phase coil, 2 — PM (mechanically unbalanced system), 3 — ferromagnetic yoke, 4 — cover, 5 — base plate, 6 — shaft, 7 — bearing, 8 — detent iron.

Market volume for household drives has increased about three times from 2000 to 2005 (Fig. 10.4).

Advances in cellular telecommunications have made mobile phones highly popular communication tools in modern society. Cell phones notify the users of an incoming call either by a ring tone or by a vibration. The speed of vibration motors is 8000 to 11 000 rpm and frequency of fundamental vibration from 133 to 183 Hz. There are two types of vibration motors: *cylindrical (bar) type* (Fig. 10.5a) and *disc (coin) type* (Fig. 10.5b). Cylindrical vibration motors are mainly used in low price bar type mobile phones and disc type vibration motors are used in relatively more expensive folder type mobile phones. Although cylindrical type motors are cheaper, they are larger in size and have weaker vibrating capability. Disc type motors are easier to mount and have stronger vibration.

Small vibration motors (Table 10.1) with diameters from 4 (cylindrical type) to 12 mm (disc type) are manufactured in very large quantities (over 700 million in 2005). Brush type PM vibration motors (Figs 10.6 and 10.7) are nowadays replaced with PMBMs (Fig. 10.8). The trends in vibration motors for mobile phones include reduced mass and size, minimized energy consumption and guaranteed stable vibration alarming at any circumstances [45].

10.2.2 Public life applications

Public life requires electric motors primarily in the following applications:

(a) heating, ventilating and air conditioning (HVAC) systems,
(b) retail bar-code readers,

(c) clocks,
(d) automatic vending machines,
(e) automatic teller machines (ATMs),
(f) ticketing machines,
(g) coin laundry machines,
(h) money changing machines,
(i) cafeteria and catering equipment,
(j) environmental control systems,
(b) amusement park equipment.

Expansion of the small motor industry is due to rapid development of consumer electronics, industrial automotive system, household electric appliances, office automation systems, communication and traffic, military equipment, electric tools, instruments, meters and electric toys.

10.2.3 Automotive applications

The number of electric motors in automotive applications is soaring. Electric motors are found in anything that has an electrical movement or solenoid function, such as window lifts, fuel pumps, mirror and headlamp adjusters, anti-lock braking systems (ABS), clutches, automatic manual transmissions, parking brakes and electric power steering. Rapid growth in electric motors is being driven by new applications such as electrically assisted power steering systems, active suspension and brake systems, electric parking brake systems, which are now spreading to more vehicle segments.

The technical solution for improved emissions, fuel consumption and higher electrical power requirements is rapidly emerging 42 V system. These higher power requirements can not be cost effectively or technically supplied by a 12 V system. Major driving forces include:

- mandated fuel economy standards;
- mandated emissions standards;
- higher electrical loads due to convenience features.

10.3 Land, sea and air transportation

10.3.1 Hybrid electric and electric vehicles

High gasoline prices, unrest in oil producing regions and concerns about global warming call for alternative powertrains, like HEVs and EVs in cars and trucks. HEVs are now at the forefront of transportation technology development (Table 10.2). HEVs combine the internal combustion engine of a conventional vehicle with the electric motor of an EV, resulting in twice the fuel economy of conventional vehicles. Superposed torque-speed curves of a combustion engine and electric motor improve the performance of HEV providing high torque at low speed and good characteristics at full speed (Fig. 10.9). The PMBM can increase the overall torque by over 50%.

Table 10.2. Hybrid electric gasoline cars.

Make	Mass kg	Number of passengers	Combustion engine	Electric motor	Battery	Max speed km/h
Honda Insight	840	2	50 kW 3-cylinder	10kW PMBM	NiMH	
Honda Civic	1240	4	71 kW 4-cylinder	15kW PMBM	NiMH	160
Toyota Prius	1255	5	57 kW 4-cylinder	50 kW PMBM	NiMH	160
Ford Escape	1425	4 to 5	99 kW 4-cylinder	70kW PMBM	NiMH	160
Mercury Mariner	1500	5	99 kW 4-cylinder	70kW PMBM	NiMH	160

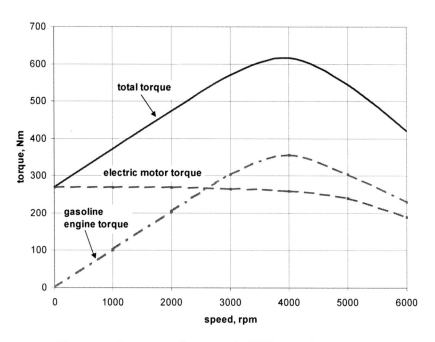

Fig. 10.9. Torque-speed curves of a HEV propulsion motors.

244   10 Scenario for nearest future

Fig. 10.10. Location of electric motor in a HEV power train. Arrow shows electric motor.

Fig. 10.11. Small high torque traction brushless motor with surface PMs (520 W, 24 V d.c., 62.1 Nm at 80 rpm). Photo courtesy *UQM Technologies*, Frederick, CO, U.S.A.

The electric motor is usually located between the combustion engine and clutch (Fig. 10.10). One end of the rotor shaft of the electric motor is bolted to the combustion engine crankshaft, while the opposite end can be bolted to

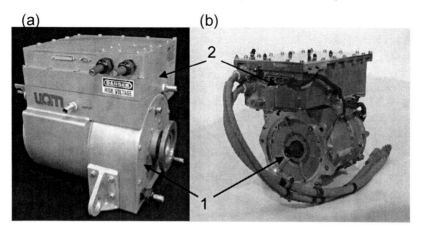

Fig. 10.12. Liquid cooled motors for HEVs and EVs integrated with liquid cooled solid state converters: (a) *UQM Technologies*; (b) *Hitachi*. 1 — motor, 2 — solid state converter.

the flywheel or gearbox via clutch (Fig. 10.10). The electric motor in a HEV serves a number of functions listed e.g., in [68].

Fig. 10.13. In-wheel PM brushless motor. Photo courtesy of TM4, Boucherville, Quebec, Canada.

Currently manufactured hybrid electric gasoline cars are equipped either with IMs or PMBMs. Industry trend is favouring PMBMs due to high torque density, high efficiency and wide constant power speed range. Some applications may still benefit from IMs because of low cost, cheap speed sensors and low windage losses. To increase the torque and power density, electric motors and solid state converters are liquid cooled, i.e., 50% of water and 50% of glycol. In most applications, the rated power of electric motors is from 10 to 75 kW. Typical PM brushless motor with surface PMs is shown in Fig. 10.11. Frequently, the electric motor is integrated with power electronics converter (Fig. 10.12).

Electric motors for HEVs and EVs can also be built in wheel. The 3-phase PM synchronous motor shown in Fig. 10.13 consists of a central stator that supports the windings and the inverter, surrounded by an external rotor which supports the PMs. The motor assembly is liquid-cooled to sustain high continuous power demands. The motor wheel uses a brushless inverted rotor configuration that can be embedded inside a regular-size wheel. The solid state converter can be installed either inside or outside of the wheel. A compact motorized wheel provides the following advantages:

- allows packaging flexibility by eliminating the central drive motor and the associated transmission and driveline components in vehicles (transmission, differential, universal joints and drive shaft);
- provides control over each wheel individually, which can result in enhanced handling and performance;
- makes it possible to regulate drive torque and braking force independently at each wheel without the need for any transmission, drive shaft or other complex mechanical components;
- produces smooth torque at low controlled speeds;
- allows for different design of vehicles.

Many road vehicles with combustion engines can be converted into EVs or HEVs. A typical HEV conversion project involves removing from the stock vehicle the engine, transmission, drive line, fuel system and exhaust system, and making the necessary modifications and system integrations to accommodate the batteries, electric propulsion system, cabling an appropriate gauges.

The primary problem with HEVs and EVs are the weight, volume, lifetime and cost of the necessary battery pack. HEVs are quiet and clean, but they can only be driven for short distance before their batteries must be recharged. On the other hand, cars with internal combustion engines can be driven anywhere to over 500 km per tank of gasoline.

10.3.2 Marine propulsion

Marine propulsion systems have been discussed in Chapter 9. Electric ship propulsion systems have several advantages over mechanical propulsion systems including increase in the useable space in a marine vessel, freedom in location of prime mover, better manoeuvrability, less vibration and noise. Electric propulsion systems also provide some level of intelligence to the power train.

The world shipping and shipbuilding industry is currently enjoying a strong upturn. Trends in engine development include electric propulsion, i.e., electric power generation, electric motors, azimuth thrusters, pod propulsors, IMPs. It has been estimated that 97% of all vessels delivered between 1999 and 2003 were powered with diesel machinery, and that 56% were direct drive, 41% were geared and 2% had an electric drive system.

10.3.3 Electric aircraft

Although propulsion of large marine vessels using low speed electric motors rated from a few to over 40 MW is now a mature technology, construction of an *all-electric passenger aircraft* is a very difficult technological challenge. So far, high speed, high frequency (1 kHz), lightweight electrical machines rated in the range of 20 to 50 MW, are not available. Fuel cell technology cannot now deliver stacks with power density minimum of 5 kW/kg, which is required for all-electric passenger aircrafts.

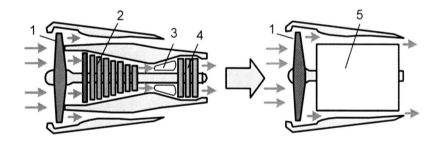

Fig. 10.14. Replacement of turbofan engine by fan electric motor. 1 — fan, 2 — compressor, 3 — combustor, 4 — turbine, 5 — electric motor.

Electric aircraft are demanded due to the following reasons:

(a) increased demand on emission and noise reduction;
(b) increased fossil fuel cost;
(c) aviation independence of oil supply;
(d) public demand for improving environmental compatibility;
(e) electric systems integration with control logic to add a level of intelligence to the aircraft power train.

Potential solutions to electric propulsion systems of aircraft using distributed fan electric motors include:

Fig. 10.15. Boeing 737 passenger aicraft consumes more than twice power than two 16-car *Shinkansen Nozomi* Series-300 bullet trains.

(a) motors powered by gas turbine driven generators;
(b) motors driven by fuel cells.

In addition, photovoltaic cells installed on wings can provide some auxiliary electric power at daytime.

Although the power density of currently available electric motors (maximum 3 kW/kg) is lower than that of turbofan engines (at least 8 kW/kg), the electric solution is much simpler, more reliable, with very low noise and no gas emissions (Fig. 10.14).

Assuming the smallest *Boeing* family passenger aircraft (B737-200) with maximum takeoff weight of 65 t, maximum fuel capacity 18.7 t and cruising speed 780 km/h, the two *Pratt & Whitney* JT8D turbofan engines provide 155 kN thrust and 26 MW power (13 MW per engine). This power requirement is equal to more than two *Nozomi Series* 300, 16-car, 1323 passenger, *Shinkansen* bullet trains (12 MW propulsion power per train). It is illustrated in Fig. 10.15. If fan electric motors are driven by gas turbine via an electric generator and power electronic converter, the following power train is required: 2×27.5 MW, 3 t each power generators; 2×27.5 MW, 1.6 t each gas turbines; 9.2 t solid state converters rated at 45 MW; two 20 MW, 2.6 t each electric motors. The total weight including cables and switchgears will be from 25 to 27 t plus weight of fuel. It has been assumed that the speed of generator is 30 krpm, speed of fan electric motors is 10 krpm with minimum efficiency 98%. Using fuel cells technology with power density 5 kW/kg (0.5 kW/kg available in 2005), the mass of the propulsion system will be reduced to 23 t. In existing propulsion systems two JT8D turbofan engines weight only 3.2 t and fuel can weight up to 18.7 t.

Thus *Boeing* or *Airbus* class electric passenger aircraft is still unreal. To build a heavy electric aircraft will require propulsion motors that are high power, lightweight and compact. Current technology cannot meet these demands because a conventional electric motor can weigh up to five times as much as conventional jet engine and not be as fuel efficient. A multidisciplinary effort is underway at the NASA Glenn Research Center to develop concepts for revolutionary, nontraditional fuel cell power and propulsion systems for aircraft applications.

It is not expected that these types of aircraft with electricity generated either by gas turbine generators or fuel cells will become a commercial products in the next 20 years. Small two-passenger aircraft have already been demonstrated and can be commercialized in the next decade.

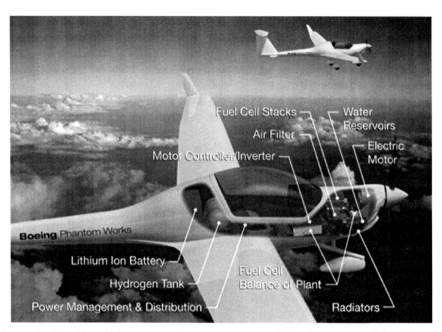

Fig. 10.16. Electrically powered two-seat motor glider demonstrated by *Boeing* [118].

In February 2008, near Madrid, Spain, *Boeing* demonstrated manned straight and level flight in a two-seat motor glider powered electrically [118]. The Boeing fuel cell demonstrator airplane uses a proton exchange membrane (PEM) fuel cell/lithium-ion battery hybrid system to power a lightweight electric motor (*UQM Technologies*, Frederick, CO, U.S.A.), which is coupled to a conventional propeller (Fig. 10.16). The fuel cell provides all power for the cruise phase of flight. During takeoff and climb, the flight segment that requires the most power, the system draws on lightweight lithium-ion batteries. The aircraft [1] has a 16.3 m wingspan, tricycle landing gear and a takeoff weight of 840 kg, including the 93 kg fuel cell (dry weight) and 10 kg of water

---
[1] Modifications have been made to Austrian *Diamond Aircraft Industries* Dimona glider.

as fuel. The cruise speed is approximately 100 km/h using fuel cell-provided power.

As a stepping stone to the all-electric aircraft, an interim solution has emerged, namely, the *More Electric Aircraft* (MEA). Such an aircraft contains some, but not all, of the key features of the all-electric aircraft. This incremental approach is attractive because it incurs significantly less risk than a wholesale change to the aircraft electrical system otherwise required [31]. Conventional aircrafts rely on bulky and complex pneumatic systems, powered by hot, high-pressure air diverted from the turbine engines. Recent advances in electric motor technology provide another opportunity to create value. For example, in the *Boeing 787 Dreamliner*, most onboard mechanical systems are powered by electric motors.

Similar power trains, as in electric aircraft, can be used for electric helicopters in which a high-speed turbine driven generator feeds a low speed propeller electric motor, e.g., TFM, via solid state converter. Electric speed reduction system is lighter than mechanical step-down gears.

Table 10.3. Specifications of SR20 UAV electric helicopter system manufactured by *Rotomotion*, Charleston, SC, U.S.A.

Length, m	1.22
Width, m	0.38
Height, m	0.560
Main rotor diameter, m	1.75
Tail rotor diameter, m	0.255
Dry weight, kg	7.5
Battery capacity, Ah	8 or 16
Output power of electric motor, kW	1.3
Climb rate, m/min	122
Maximum speed, km/h	50
Endurance, min	12 to 24 depending on battery
Maximum payload, kg	4.5

Fig. 10.17 shows a small electric helicopter operating as an *unmanned aerial vehicle* (UAV) manufactured by *Rotomotion*, Charleston, SC, U.S.A. Specifications are given in Table 10.3. The S20 UAV electric helicopter is capable of fully autonomous flight with a safety operator to perform takeoff and landing and to engage and disengage the autonomous flight control system (AFCS). It is primarily designed for all aerial photography applications. Using longitude and latitude based coordinates, the S20 UAV can repeatedly revisit the same shot locations, e.g., to show construction programs from a consistent point of view. It can operate day or night and is immune to most hazardous environments, including smoke, toxins and gunfire. It can also provide remote

Fig. 10.17. SR20 UAV electric helicopter. Photo Courtesy of *Rotomotion*, Charleston, SC, U.S.A.

inspection of projects to multiple users over the internet with complete control of the UAV and the camera.

10.4 Future trends

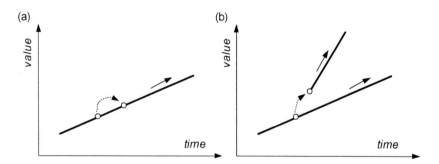

Fig. 10.18. Improvement trajectories: (a) continuous improvement; (b) discontinuous improvement.

In the late 1990s, the resurgence of American industry combined with the stagnation of the Japanese economy restored a strong emphasis on the benefit of innovations [47].

Continuous improvement (Fig. 10.18a) or continuous innovation means incremental, evolutionary improvement (fixed trajectory) or simply *doing things better. Discontinuous improvement* (Fig. 10.18b) or discontinuous innovation means breakthrough, revolutionary improvement (new steeper trajectory) or simply *doing things differently*. Discontinuous form of innovation is creation of new families of products or businesses [46]. Once a discontinuous improvement has been introduced, continuous improvement and the thinking on how to create the next discontinuous improvement must begin.

Innovation has become the industrial religion of the late 20th century [207]. Continuous improvement keeps a company in the game, while discontinuous improvement wins the game for a company. The introduction of the desktop publishing in the mid–to–late 1980s is one of the best illustrations of how a discontinuous change and innovation can in short order destroy an existing marketplace and create an entirely new playing field. Another breakthrough crown jewel is Silicon Valley.

Continuous improvement works best in slow moving industries. Electrical machine technology is very basic, but machines and the systems in which they are used are still benefiting from technological advances. Electrical machine industry can be classified as a slow moving industry. There are few exceptions like, e.g.,

(a) invention of three-phase induction motor by M. Dolivo-Dobrovolsky in 1889;
(b) invention of cylindrical rotor synchronous generator by C. Brown in 1901;
(c) impact of power electronics on control of electrical machines since the 1970s;
(d) introduction of rare-earth SmCo PMs in the 1970s and NdFEB in the 1980s;
(e) application of vector control strategy to induction and synchronous motor drives in the early 1970s;
(f) application of HTS materials in the late 1980s;
(g) application of MSM materials in the 1990s;
(h) future application of carbon nanotubes (if feasible).

These inventions have caused a breakthrough in electrical machine technology and can be counted as discontinuous improvements. While technological breakthroughs (a) to (e) and (g) have already entered the market, the future of HTS electric machines is still not clear. Can HTS electrical machines enter the market? Is their commercialization viable? Are they really competitive to high energy density PM machines?

Electrical machines operate on the principle of *electromagnetic induction law* and their physics have been unchanged since their early beginning in the 1840s. Rather development of *material engineering* than discoveries of new physical laws (unlikely) will impact the development of electrical machine technology in the future. Material engineering can also make a difference in principle of operation of electrical machines switching the physics from electromagnetic induction law to piezoelectric effect, magnetostrictive effect, MSM effect and others in the future.

Mechatronics is another trend that is increasing functionality and resulting in more precise operations by incorporating electronics directly into electrical machines.

Suppose that a new high current conductive alloy with electric conductivity twice that of copper at room temperature is invented. Application of this fictitious conductive material to electrical machines means reduction of winding losses twice or keeping the same winding losses — increase in current, current density and line current density by $\sqrt{2}$.

If a new fictitious magnetic material with saturation magnetic flux density exceeding twice the saturation magnetic flux density of silicon laminations with reduced specific core losses is invented, the magnetic flux density in the air gap of electrical machines can be increased approximately twice.

The volume of an electrical machine is inversely proportional to the line current density and air gap magnetic flux density. Applications of both two fictitious materials will bring $2\sqrt{2} \approx 2.82$ reduction of volume of an electrical machine. The mass can also be reduced approximately by the same factor. Is this a minor improvement or breakthrough? Definitely, it would be a large scale discontinuous improvement in electrical apparatus industry. The copper and silicon steels or iron-cobalt alloys would be abandoned in favor of new materials. So far, this is a science-fiction world.

Reduction of size can also be done by applying more intensive cooling systems like liquid cooling systems using oil or water and liquid gas cooling systems. The maximum current density for water and oil cooling systems is now about 30 A/mm^2 and it is very difficult to exceed this limit. With liquid gas cooling systems, on the other hand, application of SC wires becomes more economical than copper wires.

Undoubtedly, the progress in developing electrical machines is much slower than in the electronics, telecommunication or IT industries; however, this conventional technology does continue to naturally grow and evolve. Electrical machines are vital apparatus in all sectors of modern society, among them: industry, services, trade, infrastructure, healthcare, defence, and domestic life. Performing mechanical work and generating electricity, they are among the best servants of humankind and play key roles in the process of electromechanical energy conversion. Although today's electrical machines comprise more than 170 years of technology development, their technological potential has not yet been fully realized.

Abbreviations

1G	first generation HTS (multi-filament wire)
2G	second generation HTS (coated conductors)
3G	third generation HTS (conductors with enhanced pinning - future HTS wires)
a.c.	alternating current
ABS	anti-lock braking systems
A/D	analog to digital
AEW	airborne early warnings
AFCS	autonomous flight control system
AFRL	Air Force Research Laboratory
AFPM	axial flux permanent magnet (machine)
AISI	American Iron and Steel Industry
AMSC	American Superconductors (Massachussets based company)
APLC	active power line conditioner
APU	auxiliary power unit
ASIC	application specific integrated circuit
ATM	automatic teller machine
AVR	automatic voltage regulator
BEM	boundary element method
BJT	bipolar junction transistor
BPF	band pass filtering
BSCCO 2223	HTS material $Bi_{(2-x)}Pb_xSr_2Ca_2Cu_3O_{10}$
CAD	computer-aided design
CAPS	Center for Advanced Power Systems
CCD	charge coupled device
CFC	chloroflourocarbon
CHP	combined heat and power
CHPS	combat hybrid power system
CMOS	complementary metal oxide semiconductor
CNT	carbon nanotube
CPP	controllable pitch propeller

CPU	central processor unit
CSCF	constant speed constant frequency
CSD	constant speed drive
DARPA	Defence Advanced Research Project Agency
d.c.	direct current
DEW	directed energy weapon
DNA	deoxyribonucleic acid (containing genetic instructions)
DoE	Department of Energy (U.S.)
DOF	degree of freedom
DSP	digital signal processor
DSC	dynamic synchronous condenser
EAS	European Advanced Superconductors
EDM	electrical discharge machining
EMALS	electro-magnetic aircraft launch system
EMF	electromotive force
EMI	electromagnetic interference
ERAST	Environmental Research Aircraft and Sensor Technology (NASA program)
ETP	electrolytic tough pitch (copper)
EV	electric vehicle
FACTS	flexible a.c. transmission systems
FCL	fault current limiter
FDA	Food and Drug Administration (U.S.A.)
FDB	fluid dynamic bearing
FEM	finite element method
FEP	fluorinated ethylene propylene
FPC	flexible printed circuit
FPGA	field programmable gate array
FPP	fix pitch propeller
FRP	fiberglass reinforced plastic
GIEC	Guangzhou Institute of Energy Conversion (China)
GCU	generator control unit
GMTI	ground moving target indicator
GPU	ground power unit
GTO	gate turn off
HD	high definition
HDD	hard disk drive
HEL	high energy laser
HEV	hybrid electric vehicle
HPM	high power microwave
HTS	high temperature superconductor
HVAC	heating, ventilating and airconditioning
IBAD	ion beam assisted deposition
IC	integrated circuit
IDG	integrated drive generator

IFEP	integrated full electric propulsion
IGBT	insulated-gate bipolar transistor
IGCT	integrated gate commutated thyristor
IHI	Ishikawa-hajima Heavy Industry
IMP	integrated motor propeller
INCRA	International Copper Research Association
IPS	integrated power system
IPU	integrated power unit
IR	infrared radiation
ISG	integrated starter generator
IT	information technology
IVP	intracorporeal video probe
LBTS	Land-Based Test Site (Philadelphia, PA, U.S.A.)
LED	light emitting diode
LNG	liquefied natural gas
LPF	low pass filter
LVAD	left ventricular assist device
MEA	more electric aircraft
MEE	more electric engine
MEMS	microelectromechanical system
MEPS	multimegawatt electric power system
MH	metal hydride
MIS	minimally invasive surgery
MMF	magnetomotive force
MOCVD	metal organic chemical vapor deposition
MOD	metal organic deposition
MOSFET	metal oxide semiconductor (MOS) field effect transistor
MRI	magnetic resonance imaging
MSM	magnetic shape memory
MTBF	mean time between failure
MVD	magnetic voltage drop
MWNT	multi-wall nanotube
NASA	National Aeronautic and Space Administration (U.S.A.)
NdFeB	neodymium iron boron
NEDO	New Energy and Industrial Technology Development Organization (established by Japanese government)
NMR	nuclear magnetic resonance
NUWC	Naval Underwater Warfare Center
ONR	Office of Naval Research (U.S.)
ORC	organic Rankine cycle
PB	particle beam
PEM	proton exchange membrane
PFM	pulse frequency modulation
PI	proportional and integral (regulator)
PLC	programmable logic controller

PLD	programmable logic device
PM	permanent magnet
PMBM	PM brushless machine (motor)
PTI PTO	power take-in power take-out
PVD	physical vapor deposition
PWM	pulse width modulation
RABiTS	rolling assisted bi-axially textured substrate
RAT	ram air turbine
RDT	rim driven thruster, see also IMP
RFI	radio frequency interference
RFPM	radial flux permanent magnet
SAR	synthetic aperture radar
SC	superconductivity, superconductor
SEI	Sumitomo Electric Industries
SEMA	segmented electro-magnetic array
SEMP	submerged electric motor pump
SiC	silicon carbide
SMC	soft magnetic composite
SmCo	samarium cobalt
SMES	superconducting magnet energy storage system
SMPS	switching mode power supplies
SRM	switched reluctance machine (motor)
SSC	solid state converter
STO	$SrTiO_3$
SUSM	spherical ultrasonic motor
SWNT	single wall nanotube
TDF	tip driven fan
TEFC	totally enclosed fan cooled (motors)
TEWAC	totally enclosed water-air cooled
TFA	trifluoroacetate
TFM	transverse flux motor
THD	total harmonic distortion
TWT	travelling wave tube
UAV	unmanned aerial vehicle
UPS	uninterruptible power supply
USAF	United States Air Force
UUV	unmanned underwater vehicle
VCA	voice coil actuator
VF	variable frequency
VFCV	variable frequency constant voltage
VSCF	variable speed constant frequency
VSD	variable speed drives
VVVF	variable voltage variable frequency
YBCO	Yttrium Barium Copper Oxide HTS material $YBa_2Cu_3O_7$
ZVS	zero voltage switching

References

1. Accucore, TSC Ferrite International, Wadsworth, IL, USA, (2001), www.tscinternational.com
2. Adnanes A. K. (2003). Maritime electrical installations and diesel electric propulsion, ABB AS Marine and Turbocharging, Tutorial Report, Oslo, Norway. www.abb.com/marine
3. Alciatore, D. G., and Histand, M.B. (2007). *Introduction to mechatronics and measurement systems*, 3rd edition. McGraw-Hill. .
4. Alekseeva M.M. (1967). *High frequency electromechanical generators* (in Russian). Leningrad: Energia.
5. Al-Mosawi M.K., Beduz C., Goddard K., Sykulski J.K., Yang Y., Xu B., Ship K.S., Stoll R., and Stephen, N.G. (2002). Design of a 100 kVA high temperature superconducting demonstration synchronous generator, Elsevier, Physica C (372-376):1539–1542.
6. Al-Mosawi M.K., Xu B., Beduz C., Goddard K., Sykulski J.K., Yang Y., Stephen N.G. Webb M., Ship K.S., and Stoll R. (2002). 1000 kVA high temperature superconducting generator, ICEC02, pp. 237–240.
7. Amirtharajah R. , and Chandrakasan A. P. (1998). Self-powered signal processing using vibration-based power generation, IEEE Journal of Solid State Circuits, 33(5):687–695.
8. AMM Technologies, Hunfield Heights, SA, Australia, www.ammtechnologies.com
9. American Superconductors (AMSC), Westborough, MA, USA, http://www.amsuper.com/
10. Andreev E.N., Chubraeva L.I., Kunaev V.L., and Platonova M.Y. (2004). Development of high voltage superconductive alternator operating with d.c. transmission line, 6th Int. Conf. on Unconventional Electromechanical and Electr. Systems UEES04, Alushta, Ukraine, pp. 945–950.
11. Antaki J.F., Paden B.E., Burgreen G., and Groom N. (2001). Blood pump having a magnetically suspended rotor, U.S. Patent 6244835.
12. Antaki J.F., Paden B.E., Piovoso M.J., and Banda S.S. (2002). Award-winning control applications, IEEE Control System Magazine, (12):8–20.
13. Arena A., Boulougoura M., Chowdrey H.S., Dario P., Harendt C., Irion K.M., Kodogiannis V., Lenaerts B., Menciassi A., Puders R., Scherjon C., and Turgis

D. (2005). Intracorporeal Videoprobe (IVP), in *Medical and Care Compunetics 2*, edited by L. Bos et al, IOS Press, pp. 167–174.
14. Arkkio A. (1995). Asynchronous electric machine and rotor and stator for use in association therewith, U.S. Patent 5473211.
15. Arkkio A., Bianchi N., Bolognani S., Jokinen T., Luise F., and Rosu M. (2002). Design of synchronous PM motor for submerged marine propulsion systems, Int.Conf. on Electr. Machines ICEM02, Bruges, Belgium, paper No 523 (CD).
16. Arnold D.P., Zana I., Herrault F., Galle P., Park J.W., Das, S., Lang, J. H., and Allen, M. G. (2005). Optimization of a microscale, axial-flux, permanent-magnet generator, 5th Int. Workshop Micro Nanotechnology for Power Generation and Energy Conversion Apps. PowerMEMS05, Tokyo, Japan, pp. 165–168.
17. Arnold D.P., Das S., Park, J.W., Zana, I., Lang, J.H., and Allen, M.G. (2006). Design optimization of an 8-watt, microscale, axial flux permanent magnet generator, Journal of Microelectromech. Microeng., 16(9):S290–S296.
18. Atallah K., Zhu Z.Q., Howe D., and Birch T.S. (1998). Armature reaction field and winding inductances of slotless permanent-magnet brushless machines, IEEE Trans on MAG-34(5):3737–3744.
19. Auslander, D.M. (1996). What is mechatronics?, IEEE/ASME Trans. on Mechatronics, 1(1):5–9.
20. Bardeen J., Cooper, L. N., and Schrieffer J. R. (1957). Theory of superconductivity, Phys. Review, 108:1175–1204.
21. Barnes P.N. (2007). Advancing YBCO-coated conductors for use on air platforms, Int. Journal of Appl. Ceramic Technology, 4(3):242–249.
22. Bednorz J.G., and Mueller K.A. (1986). Possible high T_c superconductivity in the Ba-La-Cu-O system, Zeitschrift für Physics B – Condensed Matter, 64:189–193.
23. Beer P., Tessaro J., Eckels B., and Gaberson P. (2006). High speed motor design for gas compressor applications, 35th Turbomachinery Symposium, pp. 103–112.
24. Bent R.D.,and Mc Kinley J.L. (1981). *Aircraft electricity and electronics*, 3rd edition. McGraw-Hill Book Company.
25. Bishop R.H. (editor). (2002). *The mechatronics handbook*. Boca Raton: CRC Press.
26. Biwersi S., Billet L., Gandel P., and Prudham D. (2002). Low cost, high speed small size disc magnet synchronous motor, 8th Int. Conf. Actuator'2002, Bremen, Germany, pp. 196–200.
27. Blaschke F. (1972). The principle of field orientation as applied to the new transvektor closed loop control system for rotating field machines. Siemens Review, 39(5):217–220.
28. Blaugher R.D., Parker J.H., and McCabria J.L. (1977). High speed superconducting generator, IEEE Trans. MAG-13(1):755–758.
29. Blaugher R. D. (1994). U.S. technological competitive position, Wire Development Workshop, U.S. DoE, St. Petersburg, FL, USA.
30. Blaugher R.D. (1997). Low-calorie, high energy, IEEE Spectrum, 7:36–42.
31. Botten S.L., Whitley C.R., and King A.D. (2000). Flight control actuation technology for next generation all electric aircraft, Technology Review Journal – Millennium Issue, pp. 55–68.

32. Brady C. (2008). *The Boeing 737 technical guide.* The Boeing 737 technical site, http://www.b737.org.uk/
33. Bromberg L., Torti R., and Tekula M. (1999). Multipole magnets using monolithic hight temperature superconductor materials: I Quadrupoles. Plasma Science and Fusion Center, Report No. PSFC/RR-99-1, MIT, Cambridge, MA, USA.
34. Bruk J.S. (1928). Theory of asynchronous motor with solid rotor (in Russian), Viestnik experimentalnoy i teoretitsheskoy elektrotechniki, No. 2 and No. 5.
35. Brunvoll presents a rim driven thruster RDT, (2005), Brunvoll AS, Molde, Norway, www.brunvoll.no and www.norpropeller.no
36. Caprio M.T., Lelos V., Herbst J., and Upshaw J. (2005). Advanced induction motor end ring design features for high speed applications, Int. Electr. Machines and Drives Conf. IEMDC'05, San Antonio, TX, USA, pp. 993–998.
37. Caprio M.T., Lelos V., Herbst J., Manifold S., and Jordon H. (2006). High strength induction machine, rotor, rotor cage end ring and bar joint, rotor end ring, and related methods, U.S. Patent Publ. 2006/0273683.
38. Cascio A.M. (1997). Modeling, analysis and testing of orthotropic stator structures, Naval Symp. on Electr Machines, Newport, RI, USA, pp 91–99.
39. Ceramawire — manufacturer of ceramic coated wire, Elizabeth City, NC, USA, www.ceramawire.com
40. Chan C.C. (1996). Overview of electric vehicles — clean and energy efficient urban transportation, 7th Int. Conf. PEMC'96, Budapest, Hungary, vol. 1: K7-K15.
41. Chiba A., Fukao, T., Ichikawa, O., Oshima, M., Takemoto, M., and Dorrell, D. (2005). *Magnetic bearing and bearingless drives.* Amsterdam: Elsevier.
42. Ching N.N.H., Wong H.Y., Li W.J., Leong, P.H.W., and Wen, Z. (2002). A laser-micro machined multi-modal resonating power transducer for wireless sensing systems, Sensors and Actuators, Elsevier, vol. A97-98:685–690.
43. Choi H.Y. , Jung S.Y. and Jung H. K. (2002). Performance evaluation of permanent magnet linear generator for charging the battery of mobile apparatus, Int. Conf. On Electr. Machines ICEM02, Bruges, Belgium, paper No 234 (CD).
44. Chubraeva L.I. (2004). Prospects on power applications of high temperature superconductivity, invited paper, 6th Int. Conf. on Unconventional Electromechanical and Electr. Systems UEES04, Alushta, Ukraine, pp. 53–58.
45. Chung S.U., Hwang G.Y., Hwang S.M., Kang B.S., and Kim H.G. (2002). Development of brushless and sensorless motor used for mobile phone, IEEE Trans. MAG-38(5):3000–3002.
46. Conti T., Kondo Y., and Watson G.H. (2003). *Quality into the 21st century.* Int. Acad. for Quality.
47. Coole R.E. (1999). *Managing quality fads.* Oxford, UK: OUP.
48. Craig R.R. (1981). *Structural dynamics: an introduction to computer methods.* J. Wiley & Sons.
49. Dąbrowski M. (1977). *Construction of electrical machines* (in Polish). Warsaw: WNT.
50. Dąbrowski M., and Gieras J.F. (1977). *Induction motors with solid rotors* (in Polish). Warsaw–Poznan: PWN.
51. Day A.J., and Hirzel A. (2002). Redefining power generation, Gorham Conference, Cincinnati, OH, USA, www.lightengineering.com

52. Dev S.P. Quiet, small, lightweight, heavy fueled mini generator for power needes of soldiers and unmanned ground vehicles, D-Star Engineering Corporation, CT, USA.
53. Doyle M.R., Samuel D.J., Conway T., and Klimowski R.R. (2001). Electromagnetic aircraft launch system, Int Electr. Machines and Drives Conf. IEMDC01, Boston, MA, USA.
54. DuPont Kapton polyimide film, Motor and magnet wire industry bulletin, (2006), DuPont High Performance Films, Circleville, OH, USA, www.kapton.dupont.com
55. Eckels P.W. and Snitchler G. (2005). 5 MW high temperature superconductor ship propulsion motor design and test results, Naval Engineers Journal, Fall Issue:31–36.
56. Elektrischemachinen und Antrieb, Wintertur, Switzerland, www.eunda.ch
57. Electromechanical Corporation, Cheswick, PA, USA.
58. Felton M. (2006). Megawatt generator addresses space and electricity needs for next generation of commercial and military aircrafts, TechUpdate — A Quarterly Newsletter for MDA Technology Transfer, Summer Issue.
59. Fennimore A. M., Yuzvinsky T. D., Regan B. C., and Zettl A. (2004). Electrically driven vaporization of multiwall carbon nanotubes for rotary bearing creation, in *Electronic properties of synthetic nanostructures* edited by H. Kuzmany et al., CP723, American Institute of Physics, pp. 587–590.
60. *Film coil motor*. EmBest, Seoul, Korea, (2001), www.embest.com
61. Fingers R.T., and Rubertus C.S. (2000). Application of high temperature magnetic materials, IEEE Trans MAG-36(5):3373-3375
62. Fischer T., Potapov E.V., and Koster, A. (2003). The MicroMed deBakey VAD, Part I and II, JECT, vol. 35:274–286.
63. Flying Megawatts, Aviation Week & Space Technology, Sept 11, 2006, p. 12.
64. Fratangelo E. (2004). An integral motor/propeller for marine vessel main and auxiliary propulsion utilizing proven rim-driven technology, Curtiss-Wright Electromechanical Corporation, Cheswick, PA, USA, internal publication.
65. Friedman S. (2005). Atherectomy for plaque removal, Podiatry Management, www.podiatrym.com, June/July Issue:179–188.
66. Gertmar L. (1997). Needs for Solutions and New Areas of Applications for Power Electronics, 7th European Conf. on Power Electronics and Appl. EPE97, Trondheim, Norway, vol. 1:16–29
67. Gieras J.F., and Gieras I.A. (2001). Recent developments in electrical motors and drives, invited paper, 2nd Int. Conf. on Electrical Engineering ELECO01, Bursa, Turkey.
68. Gieras J.F., and Wing M. (2002) *Permanent magnet motors technology: design and applications*, 2ND edition. New York: Marcel Dekker.
69. Gieras J.F., and Gieras I.A. (2002). Performance analysis of a coreless permanent magnet brushless motor', IEEE 37th IAS Meeting, Pittsburgh, PA, USA, Paper No. 0-7803-7420-7/02.
70. Gieras J.F. (2003). Contemporary trends in electrical motors and drives, invited paper, 39th Int. Symp. on Electrical Machines SME03, Gdansk–Jurata, Poland.
71. Gieras J.F. (2005). Performance characteristics of a permanent magnet transverse flux generator, Int. Electr. Machines and Drives Conference IEMDC05, San Antonio, TX, USA, pp. 1293 – 1299.

72. Gieras J.F. (2006). Advancements in electric machines, Int. Conf. on Electr. Machines ICEM06. Chania, Greece, Paper No 212 (CD).
73. Gieras J.F., Oh J.H., and Huzmezan, M. (2007). Analysis and tests of an electromechanical energy harvesting oscillating device, Int. Symp. on Linear Drives for Industry Applications LDIA07, Lille, France, Paper No. 133 (CD).
74. Gieras, J.F., Wang, R.J., and Kamper, M.J. (2008) *Axial flux permanent magnet brushless machines*, 2nd edition. Dordrecht–Boston–London: Springer.
75. Glynne-Jones P., Tudor M.J., Beeby S.P., and White N.M. (2004). An electromagnetic, vibration-powered generator for intelligent sensor systems, in *Sensors and Actuators*, Elsevier, A110, pp. 344-349.
76. Graham G. (2004). Motors and motor control: high density, Design Appliance, October 1.
77. Hakala H. (2000). Integration of Motor and Hoisting Machine Changes the Elevator Business, Int. Conf. on Electr. Machines ICEM00, Espoo, Finland, 3:1242–1245.
78. Halbach K. (1980). Design of permanent multipole magnets with oriented rare-earth cobalt material, Nuclear Instruments and Methods, 169:1–10.
79. Halbach K. (1981). Physical and optical properties of rare-earth cobalt magnets, Nuclear Instruments and Methods, 187:109–117.
80. Halbach K. (1985). Applications of permanent magnets in accelerators and electron storage rings, Journal Appl Physics, 57:3605–3608.
81. Harris M.M., Jones A.C., and Alexander E.J. (2003). Miniature turbojet development at Hamilton Sundstrand: the TJ-50, TJ-120 and TJ-30 turbojets. 2nd AIAA Unmanned Unlimited Systems, Technologies and Operations – Aerospace Conference, Sand Diego, CA, U.S.A.,Paper Nr 2003-6568, pp. 1–9.
82. Hasse K. (1069). Zur Dynamik Drehzahlgeregelter Antriebe mit stromrichtergespeisten Asynchron-kurschlusslafermaschinen, PhD thesis, Technical University Darmstadt, Germany.
83. Henderson N., Kalsi S., Gamble B., and Snitchler G. (2005). Performance of a 5 MW high temperature ship propulsion Motor, Ship Technology Symposium, Biloxi, MS, USA.
84. Hoffman S., Banerjee B., and Samotyj M. (1997). Written-pole revolution, IEEE Power Eng. Review, December: 6–9.
85. Husband S.M., and Hodge C.G. (2003). The Rolls-Royce transverse flux motor development, Electrical Machines and Drives Conference IEMDC03, Madison, WI, USA, pp. 1435–1440.
86. Hussennether V., Oomen M., Legissa M., and Neumueller H.W. (2004). DC and AC properties of Bi-2223 cabled conductors designed for high-current applications, Physica C, 401:135–139.
87. Ikeda M., Sakabe S. and Higashi K. (1990). Experimental study of high speed induction motor varying rotor core construction, IEEE Trans. EC-5(1):98–103.
88. Iwakuma M., Hase Y., Satou T., Tomioka, A., Iijima Y., Saitoh T., Izumi T., Yamada Y., and Shiohara Y. (2006). Power application of YBCO superconducting tapes in Japan: transformers and motors, Int. Workshop on Coated Conductors for Applications CCA06, Ludwigsburg, Germany, 2-page abstract.
89. Iwakuma M., Tomioka A., Konno M., Hase Y., Satou T., Iijima Y., Saitoh T., Yamada Y., Izumi T., and Shiohara Y. (2007). Development of a 15 kW

motor with a fixed YBCO superconducting field winding, IEEE Trans. Appl. Superconductivity, 17(2):1607–1610.
90. Jaecklin A.A. (1997). Integration of power components — state of the art and trends, 7th European Conf. on Power Electronics and Appl. EPE97, Trondheim, Norway, 1:1–6.
91. Jansson P. (1999). Soft magnetic composites — a rapidly expanding materials group, Int. Conf. on Powder Matallurgy and Particulate Materials, Vancouver, Canada.
92. Jintanawan T., Shen I.Y., and Tanaka K. (2001). Vibration analysis of fluid dynamic bearing spindles with rotating-shaft design, IEEE Trans. MAG-37(2):799–804.
93. Kalsi S. (2003). A small-size superconducting generator concept, Int. Electr. Machines and Drives Conf., IEMDC03, Madison, WI, USA, pp. 24–28.
94. Kennett R.J. (1981). Advanced generating system technology, Aicraft Electrical Power Systems SP-500 Conf., SAE, Warrendale, PA, USA, pp. 35–41.
95. Kessinger R.L., and Robinson S. (1997). SEMA-based permanent magnet motors for high-torque, high-performance, Naval Symp. on Electr. Machines, Newport, RI, USA, pp. 151–155.
96. Kessinger R.L., Stahura P.A., Receveur P.E., and Dockstader K.D. (1998). Interlocking segmented coil array, U.S. Patent 5744896.
97. Kessinger R.L. (2002). Introduction to SEMA motor technology, Kinetic Art and Technology, Greenville, IN, USA.
98. Kimmich R., Doppelbauer M., Kirtley J.L., Peters D.T., Cowie J.G., and Brush E.F. (2005). Performance characteristics of drive motors optimized for die-cast copper cages, Energy Efficiency in Motor Driven Systems EEMODS05, Heidelberg, Germany, pp. 1–8.
99. King J., and Ritchey I. (2005). Marine propulsion: The transport technology for the 21st century?, Ingenia, pp. 7–14.
100. Kitano Seiki Co., Ltd., www.kitano-seiki.co.jp
101. Kleen S., Ehrfeld W., Michel F., Nienhaus M., and Stölting H.D. (2000). Penny-motor: a family of novel ultraflat electromagnetic micromotors, Int. Conf. Actuator2000, Bremen, Germany, pp. 193–196.
102. Kleiner F. and Kauffman, S. (2005). All electric driven refrigeration compressors in LNG plans offer advantages, Gastech2005, Bilbao, Spain, pp. 1–5.
103. Kode V.R.C., and Cavusoglu M.C. (2007). Design and characterization of a novel hybrid actuator using shape memory alloy and d.c. micromotor for minimally invasive surgery applications, IEEE/ASME Trans. Mechatronics, 12(4):455-464.
104. Kolowrotkiewicz J., Baranski M., Szelag W., and Dlugiewicz L. (2007). FE analysis of induction motor working in cryogenic temperature, Compel – the Int. Journal for Computations and Mathematics in Elec. Eng., 26(4): 952–964.
105. Kornbluh R., Pelrine J., Eckerle J., and Joseph, J. (1998). Electrostrictive polymer artificial muscle actuators, Int. Conf. on Robotics and Automation ICRA98, pp. 2147–2154.
106. Krovel O., Nilssen R., Skaar S. E., Lovli E., and Sandoy N. (2004). Design of an integrated 100 kW permanent magnet synchronous machine in a prototype thruster for ship propulsion, Int. Conf. on Electr. Machines ICEM04, Krakow, Poland, Paper No. 697 (CD).

107. Krovel O., Nilssen R., and Nysveen A. (2004). Study of the research activity in the nordic countries on large permanent magnet synchronous machines, Nordic Workshop on Power and Industr Electronics NORPIE04, Trondheim, Norway, Paper No. 053.
108. Kubzdela S., and Węgliński B. (1988). Magnetodielectrics in induction motors with disk rotor, IEEE Trans MAG-24(1):635–638.
109. Kyura N., and Oho H. (1996). Mechatronics — an industrial perspective, EEE/ASME Trans. Mechatronics, 1(1):10–15.
110. Lange A., Canders W.R., Laube F., and Mosebach H. (2000). Comparison of different drive systems for a 75 kW electrical vehicle drive, Int. Conf. on Electr. Machines ICEM00, Espoo, Finland, 2:1308-1312.
111. Lave L.B., Hendrickson C., and McMichael F.C. (1994). Recycling decisions and green design, Environmental, Science and Technology, 28(1):18A–24A.
112. Lee J.M.H., Yuen S.C.L., Luk M.H.M., Chan G.M.H., Lei K.F., Li W.J., Leong P.H.W, and Yam Y. (2003). Vibration-to-electrical power conversion using high-aspect-ratio MEMS resonators, Power MEMS Conf., Chiba, Japan.
113. Liebermann E. (2003). Rotor cooling arrangement, U.S. Patent 6661133.
114. Lieutaud P., Brissonneau P., Chillet C., and Foggia A. (1991). Preliminary investigations in high speed electrical machines design, Int. Conf. on the Evolution and Modern Aspects of Synchronous Machines SM100, Zurich, Switzerland, 3:840–844.
115. Mangan J., and Warner A. (1998). Magnet wire bonding, Joyal Product, Inc., Linden, NJ, USA, www.joyalusa.com
116. Mashimo T., and Toyama S. (2007). MRI-compatibility of a manipulator using a SUSM, 12th IFToMM World Congress, Besancon, France.
117. Matsuoka K., Obata S., Kita H., and Toujou F. (2001). Development of FDB spindle motors for HDD use, IEEE Trans. MAG-37(2):783–788.
118. Mecham M. (2008). Boeing complets fuel cell flights, Aviation Week, April 3, www.aviationweek.com
119. Meninger S., Mur-Miranda J.O., Amirtharajah R., Chandrakasan A.P., and Lang H.J. (2001). Vibration–to–electric energy conversion, IEEE Trans. VLSI Systems, 9(1):64–76.
120. Micro power generation based on micro gas turbines: a challenge. KU Leuven, Belgium. http://www.powermems.be
121. Miniature motors. Portescap^{TM}— A Danaher Motion Company, La Chaux-de-Fonds, Switzerland, 2006, www.portescap.com
122. Mishler W.R. (1981). Test results on a low amorphous iron induction motor. IEEE Trans. PAS-100(6):860-866.
123. Mongeau P. (1997). High torque/high power density permanent magnet motors, Naval Symp. on Electr. Machines, Newport, RI, USA, pp. 9–16.
124. Morash R.T., Barber R.J., and Roesel J.F. (1993). Written-pole motor: a new a.c. motor technology, Int. Conf. on Electr. Machines in Australia, ICEMA93, Adelaide, Australia, pp. 379–384.
125. Morita G., Nakamura T., and Muta I. (2006). Theoretical analysis of a YBCO squirrel-cage type induction motor based on an equivalent circuit, Superconductor Science and Techn., 19:473–478.
126. Nagao K., Nakamura T., Nishimura, T., Ogama, Y., Kashima N., Nagaya S., Suzuki K., Izumi T., and Shiohara Y. (2008). Development and fundamental characteristics of YBCO superconducting induction-synchronous

motor operated in liquid nitrogen, Superconductor Science and Techn., 21(1):015022(5pp).
127. Nanocrystalline cores with high permeability and lowe core loss, Guangdong Coilcore Technology Development, Guangzhou City, China, www.coilcore.com
128. Nakamura T., Miyake H., Ogama Y., Morita G., Muta I., and Hoshino T. (2006). Fabrication and characteristics of HTS induction motor by the use of Bi-2223/Ag squirrel-cage rotor, IEEE Trans. Applied Superconductivity, 16(2):1469–1472.
129. Nakamura T., Ogama Y., and Miyake H. (2007). Performance of inverter fed HTS induction-synchronous motor operated in liquid nitrogen, IEEE Trans. Applied Superconductivity, 17(2):1615–1618.
130. Nakamura T., Ogama Y., Miyake H., Nagao, K., and Nishimura T. (2007). Novel rotating characteristics of a squirrel-cage-type HTS induction/synchronous motor, Superconductor Science and Technology, 20(6):911–918.
131. Nanocrystalline *Vitropen* EMC components, Vacuumschmelze, Hanau, Germany, www.vacuumschmelze.com
132. Navigant Consulting, Burlington, MA, USA, www.navigantconsulting.com
133. Nayak, O., Santoso, S. and Buchanan, P. (2002). Power electronics spark: new simulation challenges, IEEE Computer Applications in Power, No 10, pp. 37–44.
134. Neshkov T., and Dobrinov A. (2006). Novel type grippers for small objects manipulation, Problems of Engineering, Cybernetics and Robotics, 57, BAS, Sophia, Bulgaria, pp. 44–52.
135. Neumüller H.W., Nick W., Wacker B., Frank M., Nerowski G., Frauenhofer J., Rzadki W., and Hartig R. (2006). Advances in and prospects for development of high-temperature superconductor rotating machines at Siemens, Superconductor Science and Techn., 19:114-117.
136. Neumüller H.W., Klaus G., and Nick W. (2006). Status and prospects of HTS synchronous machines, Int. Conf. on Modern Materials and Techn. CIMTEC06, Acireale, Sicily, Italy.
137. Nexans — global expert in cables and cabling systems, Paris, France, www.nexans.com
138. Nichols S., Foshage J., and Lovelace E. (2006). Integrated motor, propulsor (IMP) and drive electronics for high power density propulsion, Electric Machine Technology Symposium EMTS06, Philadelphia, PA, USA.
139. NKG Berylco, NGK Metals Corporation, Sweetwater, U.S.A., www.ngkmetals.com
140. Nonaka S., and Oomoto M. (1998). A brushless 4-pole three-phase synchronous generator with cylindrical rotor, Int. Conf. on Electr. Machines ICEM98, Istanbul, Turkey, 2:1362–1367.
141. Nord G., Jansson P., Petersen C.C., and Yamada T. (2005). Vertical electrical motor using soft magnetic composites, Int. Electr. Machines and Drives Conf. IEMDC05, San Antonio, TX, USA, pp. 373–377.
142. Okazaki T., Hayashi K., and Sato K. (2006). Industrial application of HTS coils using next-generation BSCCO Wire, SEI Technical Review, No. 61, pp. 24–28.
143. O'Neil S.J. (1997). Advances in motor technology for the medical industry, Medical Device and Diagnostic Ind. Magazine, May issue.

144. Pallett E.H.J. (1998). *Aircraft electrical systems*, 3rd edition, 9th impression. Harlow: Addison Wesley Longman.
145. Parker J.H., Blaugher R.D., Patterson A., and McCabria, J.L. (1975). A high speed superconducting rotor, IEEE Trans. MAG-11(2):640–644.
146. Patterson D., and Spee R. (1995). The design and development of an axial flux permanent magnet brushless d.c. motor for wheel drive in solar powered vehicles, IEEE Trans IA-31(5):1054–1061.
147. Pazik J.C. (2006). ONR science and technology for integrated systems approach for the all electric force, keynote presentation, Electric Machine Technology Symp. EMTS 2006, Philadelphia, PA, USA.
148. Plaque excision in the peripheral, Supplement to Endovascular Today, September 2004, pp. 1 - 11.
149. Powerfull electrical solutions in oil and gas industry, Converteam, Messy, France, (2007), www.converteam.com
150. Properties of MSM materials, AdaptaMat, Helsinki, Finland, www.adaptamat.com
151. Puttgen H.B., MacGregor P.R., and Lambert F.C. (2003). Distributed generation: semantic hype or the dawn of a new era?, IEEE Power and Energy Magazine, 1(1):22–29.
152. Reichert K., and Pasquarella G.: Highs speed electric machines, status, trends and problems, invited paper, IEEE/KTH Stockholm Power Tech Conf., Stockholm, Sweden, 1995, pp. 41–49.
153. Radun A.V., and Richter E. (1993). A detailed power inverter design for a 250 kW switched reluctance aicraft engine starter/generator, SAE Aerospace Atlantic Conference and Exposition, Dayton, OH, USA, Paper No. 931388, pp. 274–288.
154. Radun A.V., Ferreira C.A., and Richter E. (1998). Two-channel switched reluctance starter–generator results, IEEE Trans. IA-34(5):1026–1034.
155. Rahim Y.H.A., Mohamadien A.L., and Al Khalaf A.S. (1990). Comparison between the steady-state performance of self-excited reluctance and induction generators, IEEE Trans. EC, 5(3):519–525.
156. Ramsden V.S., Mecrow B.C., and Lovatt H.C. (1997). Design of an in-wheel motor for a solar-powered electric vehicle, Int Conf. on Electr. Machines and Drives EMD97, IEE, London, pp. 192–197.
157. Ramdsden V.S., Watterson P.A., Holliday W.M., Tansley G.D., Reizes J.A., and Woodard J.C. (2000). A rotary blood pump, Journal IEEE Australia, 20:17–22.
158. Rare earth permanent magnets, Vacuumschmeltze, Hanau, Germany, (2007), www.vacuumschmeltze.de
159. Research, development and manufacture in high tech areas, *Oswald Elektromotoren*, Miltenberg, Germany, www.oswald.de
160. RF System Lab, Nagano-shi, Japan, www.rfsystemlab.com
161. Rich, D.: The no-cog motor. (2006). Machine Design, May 11.
162. Rolls-Royce marine propulsion products, London, UK, (2004), www.rolls-royce.com/marine/products/default.jsp
163. Rush S. (2004). Submerged motor LNG pumps in send-out system service, Pumps & Systems, May Issue:32–37, www.pump-zone.com
164. Salle D., Cepolina F., and Bidaud P. (2004). Surgery grippers for minimally invasive heart surgery, IEEE Int. Conf. on Inttell. Manip. and Grasping IMG04, Genova, Italy.

165. SatCon Corporation, SatCon Power Systems, West Boylston, MA, USA, www.satcon.com
166. Scherzinger W.M., Chu T., and Kasdan L.M. (1990). Liquid cooled salient pole rotor support wedges, U.S. Patent 4943746.
167. Schiferl R. (2004). Development of ultra efficient HTS electric motor systems, 2004 Annual Superconductivity Peer Review Meeting, Washington DC, USA.
168. Schob R., and Bichsel J. (1994). Vector control of the bearingless motor, 4th Int. Symp. on Magn. Bearings, Zurich, Switzerland, pp. 327–352.
169. Secunde R.R., Macosko R.P., and Repas D.S. (1972). Integrated engine-generator concept for aicraft electric secondary power, Report No. NASA TM X-2579, NASA Lewis Research Center, Cleveland, OH, USA.
170. Segrest J.D., and Cloud W.W. (1981). Evolution and development of high voltage (270 V) d.c. aircraft electric systems in the United States, Aicraft Electrical Power Systems SP-500, SAE, Warrendale, PA, USA, pp. 51–63.
171. Servax drives, Landert-Motoren AG, Bulach, Switzerland, www.servax.com
172. Shearwood C., and Yates R. B. (1997). Development of an electromagnetic microgenerator, Electronics Letters, 33(22):1883–1884.
173. Shenck N.S., and Paradiso J.A. (2001). Energy scavenging with shoe-mounted piezoelectrics, IEEE Micro, 21(3):30–42.
174. Shenfier K.I. (1926). Rotor of asynchronous motor in form of a solid iron cylinder (in Russian), Elektritshestvo, No. 2, pp. 86–90.
175. Sivasubramaniam K., Laskaris E.T., Shah M.R., Bray J.W., and Garrigan N.R. (2006). HTS HIA generator and motor for naval applications, Electric Machine Technology Symposium EMTS 2006, Philadelphia, PA, USA.
176. Small motors bulks up. (1997). Machine Design, May 22, p. 36. See also. (2000). Smart Motor – integrated motor controller amplifier, Animatics Co., Santa Clara, CA, USA, www.animatics.com
177. Soft magnetic cobalt-iron alloys, PHT-004, Vacuumschmelze, Hanau, Germany, 2001, www.vacuumschmelze.de
178. Southall H.L., and Oberly C.E. (1979). System considerations for airborne, high power superconducting generators, IEEE Trans. MAG-15(1):711–714.
179. Stec T.F. (1994). Amorphous magnetic materials Metglass 2605S-2 and 2605TCA in application to rotating electrical machines, NATO ASI Modern Electrical Drives, Antalya, Turkey.
180. Stec T. (1995). Electric motors from amorphous magnetic materials, Int. Symp. on Nonlinear Electromagnetic Systems, University of Cardiff, Cardiff, UK.
181. Stefani P., and Zandla G. (1992). Cruise liners diesel electric propulsion. Cyclo or synchroconverter? The shipyard opinion, Int. Symp. on Ship and Shipping Research, Genoa, Italy, pp. 2:6.5.1–6.5.32
182. Steimer C.K., Grüning H.E., Werninger J., Caroll E., Klaka S., and Linder S. (1999). IGCT — a new emerging technology for higher power, low cost converters, IEEE IA Magazine 5(4):12–15.
183. Sterken T., Baert K., Puers R., and Borghs S. (2003). Power extraction from ambient vibration, Catholic University of Leuven, Leuven, Belgium.
184. Stevens S., Deliege G, Driesen J., and Belmans, R. (2005). A hybrid high speed electrical micromachine for microscale power generation, Catholic University of Leuven, Leuven, Belgium.

185. Stoianovici D., Cadeddu J., Demaree R.D., Basile H.A., Taylor R.H., Whitcom L.L., and Kavoussi L.R. (1997). A novel mechanical transmission applied to percutaneous renal access, ASME Dynamic and Control Division, 61:401–406.
186. Stoianovici D., Patriciu, A., Mazilu, D., and Kavoussi L. (2007). A new type motor: pneumatic step motor, IEEE/ASME Trans. Mechatronics, 12(2):98–106.
187. Sugimoto H., Nishikawa T., Tsuda T., Hondou Y., Akita Y., Takeda T., Okazaki T., Ohashi S., and Yoshida Y. (2006). Trial manufacture of liquid nitrogen cooling high temperature superconductivity rotor, Journal of Physics: Conference Series, 43:780–783.
188. SuperPower, Schenectady, NY, USA, www.superpower-inc.com
189. Swenski D.F., Norman L.W., and Meshew, A.D. (1972). Advanced airborne auxiliary power system, 39th Meeting of the AGARD propulsion and Energetics Panel, USAF Academy, Colorado Springs, CO, USA, pp. 17.1–17.12.
190. Takeda T., Togawa H., and Oota, T. (2006). Development of liquid nitrogen-cooled full superconducting motor, IHI Engineering Review, 39(2):89–94.
191. Tanaka H., and Kuga T. (1985). 2000 kW, 8000 rpm very high speed induction motor drive system, IEEE IAS Meeting, pp. 676–680.
192. Technology transfer report — the die-cast copper motor rotor, Copper Development Association Inc., New York, USA, 2004, Revision No 4.
193. Tests procedures for synchronous machines, IEEE Guide, Standard 115-1995 (R2002).
194. The smallest drive system in the world, *Faulhaber*, Schonaich, Germany, (2004), www.faulhaber.com
195. Thin non oriented electric steels, Cogent Power Ltd, Newport, UK, (2005), www.cogent-power.com
196. Tolbert L.M., Peterson W.A., Theiss T.J., and Scudiere M.B. (2003). Gensets, IEEE IA Magazine, 9(2):48–53.
197. Tolliver J., Rhoads G., Barnes P., Adams S., and Oberly C. (2003). Superconducting generators: enabling airborne directed energy weapons, 1st Int. Energy Conversion Eng. Conf. IECEC03, Portsmouth, VA, USA.
198. Tomkinson D., and Horn J. (1995). *Mechatronics engineering*. New York: McGrow-Hill.
199. Torah R.N., Beeby S.P., Tudor M.J., O'Donnell T., and Roy S.: Development of a cantilever beam generator employing vibration energy harvesting, University of Southampton, UK.
200. Tsai M.C., and Hsu, L.Y. (2007). Winding design and fabrication of a miniature axial-flux motor by micro-electroforming, IEEE Trans. MAG-43 (7):3223–3228.
201. Turowski J. (2003). Mechatronics and electrical machines, invited paper, 39th Int. Symp. on Electr. Machines SME03, Gdansk–Jurata, Poland, pp. 1–8.
202. Turowski J. (2004). Innovative challanges in technology management, chapter in book *Transition in the European research and innovation area* edited by Jasinski A.H., University of Warsaw, Warsaw, pp. 192–211.
203. Turowski J. (2008). *Fundamentals of mechatronics* (in Polish). Lodz, Poland: Higher School of Humanities and Information Technology.
204. Ueha S., Tomikawa Y., Kurosawa M., and Nakamura N. (1993). *Ultrasonic motors. Theory and applications*. Oxford, UK: Clarendon Press.

205. Ullakko K. (1996). Magnetically controlled shape memory alloys: a new class of actuator materials, Journal of Material Eng. and Performance, 5:405–409.
206. Umans S.D., and Shoykhet B. (2005). Quench in high-temperature superconducting motor field coils: experimental resuls, IEEE IAS Conf. and Meeting, Hong Kong, pp. 1561–1568.
207. Valery N. (1999). Innovation in industry. Industry gets religion. The Economist. A survey of innovation in industry, February 20, pp. 5–8.
208. Van Buijtenen J.P., Larjola J., Turunen-Saaresti T., Honkatukia J., Esa H., Backman J., and Reunanen A. (2003). Design and validation of a new high expansion ratio radial turbine for ORC application, 5th European Conf. on Turbomachinery, Prague, Republic of Czech, pp. 1091–1104.
209. Vas P. (1999). *Sensorless vector and direct torque control* Oxford, UK: Clarendon Press.
210. Węgliński B. (1990). Soft magnetic powder composites — dielectromagnetics and magnetodielectrics, Reviews on Powder Metallurgy and Physical Ceramics, London: Freund Publ. House, 4(2).
211. Williams C.B., and Yates R. B. (1995). Analysis of a micro-electric generator for microsystems, 8th Intern. Conf. on Solid-State Sensors and Actuators and Eurosensors IX, Stockholm, Sweden, pp. 369–372.
212. Wolfgang E., Niedemostheide F.J., Reznik D., and Schulze, H.J. (1999). Advances in power electronics devices, Int. Electr. Machines and Drives Conf. IEMDC99, Seattle, WA, USA, pp. 4–8.
213. Written-pole technology, Precise Power Corporation, Palmetto, FL, USA, www.precisepwr.com
214. Zordan M., Vas P., Rashed M., Bolognani S., and Zigliotto M. (2000). Field-weakening in high-performance PMSM drives: a comparative analysis, IEEE IA Conf. and Meeting, Rome, Italy, pp. 1718–1724.
215. Zwyssig C., Kolar J.W., Thaler W., and Vohrer M. (2005). Design of a 100 W, 500 000 rpm permanent magnet generator for mesoscale gas turbines, IEEE 40th IAS Conf. and Annual Meeting, Hong Kong, 1:253–260.

Index

actuation technologies, 136
actuator
   for medical applications, 135
   high power, 156
   linear, 3, 142
   locally installed, 155
   magnetostrictive, 136, 153
   MSM, 68, 153
   piezoelectric, 64, 136, 153
   pneumatic, 153
   short-stroke, 142
   small size, 155
   tendon-type, 155
   thermal expansion, 153
   VCA, 136, 152
airborne
   AEW systems, 98
   apparatus, 2, 125
   applications, 202
   military systems, 81
   power mission, 81
   radar, 81, 97, 202
aircraft
   Airbus, 248
   all-electric, 247, 250
   Boeing, 248, 250
   commercial, 97
   Dimona, 249
   electric helicopter, 250
   electric propulsion, 247
   military, 97
   more electric, 2, 250
   power train, 125
   powered electrically, 249
   solar powered, 124
amorphous alloys, 36
amplifier, 128
automotive applications, 242
azimuthing thruster, 214, 219, 220, 222, 247

ball bearing, 138, 237
ball screws, 3
battery, 129, 135, 143, 147, 157, 167, 246, 249
beryllium copper, 108
BSC theory, 17

cam, 151
capsule endoscopy, 148, 150, 151
catheter, 136, 146, 147
cavitation, 220
CHP, 87, 162
cobalt alloys
   contents, 31
   Hiperco, 31, 163
   Vacoflux, 33
colonoscopy, 148
compressor, 79, 86, 135, 159, 161
   variable speed, 89
   with SRM, 89
   with PMBM, 88
control
   circuitry, 21
   closed loop, 139
   computer, 155
   direct, 12

field oriented, 12
indirect, 12
microprocessor, 139
motion, 3
of generator, 92
open loop, 139
pitch, 219
precision, 151
process, 129
remote, 155
strategies, 2
system, 139
technology, 5
converter, 11, 86, 95, 157, 218, 226
  integrated, 246
  liquid cooled, 246
  operation, 187
  water cooled, 21
cooling
  active, 222
  air, 21, 72, 104
  air-air, 79
  convection, 205
  liquid helium, 51
  medium, 84, 99, 207
  oil, 160
  sea water, 222
  spray oil, 96, 102
  techniques, 102
  water, 21
  water-air, 79
  water-glycol, 246
cooling system
  air-water, 187
  cold plate, 93, 99, 226
  comparison, 102
  cryogenic, 176, 183
  cryostat, 190, 198, 207
  direct, 85, 102
  effective, 99
  efficient, 84
  forced air, 96
  forced oil, 96
  liquid, 7, 76, 84, 206, 246
  liquid gas, 253
  liquid helium, 205
  liquid jacket, 93, 104
  liquid nitrogen, 171, 190
  microturbine, 84
  more intensive, 253
  of the rotor, 93
  thermosyphon, 185
  water, 72
  with cryocooler, 186
Cooper pairs, 17
copper cage, 108, 132, 133
core
  slotless, 41, 143, 173
  slotted, 41
critical
  current, 52, 191, 193
  current density, 14, 52
  field, 15
  magnetic field, 14
  state, 61
  temperature, 14, 19
cycle
  Bryton, 86
  combined, 87
  organic Rankine, 87, 111
  Rankine, 87

da Vinci robot, 153, 155
DARPA, 18
degree of freedom, 143, 152, 154, 155
device
  CCD, 154
  clinical engineering, 136
  comodity, 135
  electromagnetic, 27
  electromechanical, 129
  energy harvesting, 167, 169
  LVAD, 140, 144, 156
  medical, 135, 156
  SilverHawk, 147
  solid state, 213
  surgical, 139, 156
directed energy weapons, 2, 81, 96, 202
disc type, 123
DoE, 18
drive
  compact, 131
  direct, 130
  electromechanical, 130
  integrated, 2, 130
  ISG, 130
  large, 191
  linear, 69

podded, 171
shaft, 215
variable frequency, 179
variable speed, 2, 21, 79, 213, 218
vector controlled, 11
DSC, 185
dynamometer, 179

effect
cogging, 156
deep bar, 132
double cage, 132
Josephson, 14
magnetostrictive, 252
Meissner, 14, 52
MSM, 252
piezoelectric, 252
skin, 132
electric ship, 213, 217
electrical steels
grading, 28
nonoriented, 27
thin gauges, 29
electricity consumption, 22
electronics, 13, 125, 233, 253
elevator
gearless, 76
space, 62
encoder, 75, 131
end-effector, 152, 155, 156
endoscope, 148, 154
energy
applications, 209
conservation, 21
consumption, 21, 159, 191
conversion, 213, 253
density, 7, 156, 157
electrical, 23
generated, 169
harvesting, 165, 169
kinetic, 166, 168
renewable, 235
saving, 2
thermal, 87
entropy, 87
EV, 2, 242, 246

fault tolerance, 83, 156
FCL, 211

field orientation, 11
flux pinning, 60
flux trapping, 58
flywheel, 129, 130
frequency
high, 75, 81, 92
natural, 169
of vibration, 169, 241
output, 95
switching, 11
utility, 75
fuel cell, 125, 247–249

generator
air core, 185
airborne, 99
aircraft, 89, 92, 95, 96
APU, 89
brushless, 83, 160
conventional, 129
coreless, 184
cryogenic, 81
CSCF, 95
direct drive, 104
disc type, 163
dual channel, 96
engine driven, 89
for soldiers, 157
GPU, 90
high speed, 81, 82, 96, 104, 111, 207
homopolar, 81, 200
HTS, 183, 187, 198, 205
HTS high speed, 97, 202
HTS homopolar, 200, 206, 217
HTS synchronous, 81, 97, 171, 217, 223
ironless, 160
ISG, 104, 129
kinetic, 165
large, 172
lightweight, 99
linear, 167
LTS high speed, 202
main, 95
megawatt-class, 99
micro, 157, 163, 164
mini, 157, 160, 161
miniature, 157, 162, 167
modern, 83

moving magnet, 167
MSM, 167
multimegawatt, 96
of magnetic field, 151
piezoelectric, 167
PM, 95
PM brushless, 92
RAT, 89
reluctance, 162
SC, 183, 184
slotless, 162
switched reluctance, 92, 96
synchronous, 92, 186, 213, 217, 223
thermoelectric, 165
VF, 95
vibration, 167
voltage regulated, 92
VSCF, 95
wind, 126
written pole, 117
gripper, 69, 122, 135, 152–156
GTO, 19
guided capsule, 151

Halbach array, 47, 77, 124, 232
handpiece, 122, 135
HDD, 235
  FDB, 237
  tied, 237
  untied, 237
heat exchanger, 85, 99, 222
HEV, 2, 242, 245, 246
high power density, 71, 83
high power microwave, 96, 213
hydrodynamic bearing, 138

IC, 130, 165
IDG, 95
IGBT, 19, 21, 61, 234
IGCT, 21
IMP, 221, 222, 229, 231, 247
impeller, 142–144
implantable, 135
improvement
  continuous, 251
  discontinuous, 252
  incremental, 251
  large scale, 253
inductive power transfer, 151

inductor, 128
innovation, 251
  continuous, 251
  discontinuous, 252
insulation, 48
  ceramic, 50
  class, 102
  heat activated, 48
  high temperature, 50
  Kapton, 50
  thermoplastic, 205
integrated
  devices, 13
  electric propulsion, 213
  electromechanical drives, 2
  impeller, 144
  motor, 144, 246
  motor-propeller, 221
  power system, 224
  rotor, 237
integration, 3, 156, 159
inverter, 11, 89, 95, 118, 120, 162

klystron, 98

laparoscope, 151, 154
laser, 96
lattice, 16
law
  electromagnetic induction, 252
  Moore's, 165
levitation, 16, 19, 143
Liberty of the Seas, 221
LNG, 5, 79
loading
  electric, 71, 253
  magnetic, 71, 253
LVAD, 141, 144, 156

machines
  applications, 81, 111
  breaktrough, 252, 253
  classical, 185
  development, 235, 253
  electrical, 12, 23, 27, 81
  for compressor, 104
  gearless, 81
  high power density, 44
  high speed, 81, 102, 111

homopolar, 19, 200
HTS, 185, 252
HTS high speed, 207
HTS low speed, 171, 207
HTS synchronous, 172
hybrid reluctance, 159
induction, 103, 104, 109
lightweight, 247
nano-electromechanical, 62
progress, 12
recyclable, 25
switched reluctance, 96
synchronous, 19, 91
trends, 1, 251
with cage rotor, 103
with solid rotor, 109, 111
written pole, 115
magnet
  inserted in cavity, 124
  quadruple, 58
magnetic bearings, 2, 81, 104, 120, 138, 230
magnetic shape memory (MSM), 68
magnetization curve, 28, 29, 33
magnetostriction, 68
market, 3, 8, 209, 211, 221, 241, 252
material engineering, 1, 27, 252
materials
  bulk HTS, 58
  ceramic, 50
  fictitious, 253
  high temperature, 49, 132
  impregnating, 50
  magnetostrictive, 68
  MSM, 68
  nanocrystalline, 62, 67
  nanostructured, 62
  new, 132
  nonmagnetic, 155
  requirements, 139
  SC, 184
mechatronics, 2, 13, 253
MEMS, 13
microturbine, 82, 84, 86, 111, 157, 159
missiles, 162
mobile phones, 241
moment of inertia, 216, 235
MOSFET, 19, 61
motor

1.9 mm, 147
13 mm, 136, 153
4 mm, 151
advanced, 224, 225
AFPM, 43, 144
applications, 118
auxiliary, 129
axial flux, 226
bearingless, 2, 119
brushless, 144
comparison, 4, 11, 225
converter fed, 23
coreless, 123, 124, 128, 135, 147, 156
disc type, 76, 218, 226, 241
energy efficient, 1, 23
for aircraft, 247
for capsule, 150
for cell phone, 241
for compressor, 79, 88, 111
for cooling fan, 235
for EV, 242
for HDD, 121, 235
for HEV, 242
for household, 238
for LNG plants, 79
for medical applications, 135
for pump, 111, 121
for ship propulsion, 225, 234
gearless, 2
high power, 156
high speed, 104, 105, 113
homopolar, 199
HTS, 189, 191, 234
HTS 36.5 MW, 180
HTS 5 MW, 174
HTS axial flux, 191
HTS disc type, 189, 196
HTS homopolar, 200
HTS induction, 192
HTS synchronous, 171, 173, 194
IMP, 229, 231
in wheel, 246
induction, 1, 4, 11, 105, 113, 132
integrated, 144
iron free, 123
large, 5, 78, 173, 216, 224, 225
lightweight, 125
linear, 142, 152
locally installed, 155

MEMS, 14
modern, 135, 156
moving magnet, 126
nano, 65
piezoelectric, 118, 119
PMBM, 1, 7, 76, 131, 138, 142, 152, 239
PMBM 36 MW, 225
PneuStep, 155
propulsion, 217, 218
Q-PEM, 151
SC, 173
SC reluctance, 198
servo, 11, 111
ship propulsion, 171, 173
single-phase, 117
slotless, 121, 122, 126, 128, 144, 156
small, 242
spindle, 237
square wave, 1
SRM, 10, 239
stepping, 9, 150, 151
submerged, 220
SUSM, 155
synchronous, 1, 5, 79, 116, 208, 225, 234, 246
test facility, 178
TEWAC, 79
three phase, 246
tiny, 150
transverse flux, 71, 75, 218, 228
ultrasonic, 118, 153, 155
variable speed, 218
very small, 135
vibration, 241
with cryogenic cooling, 79
with gearhead, 146
with solid rotor, 109
wobble, 151
wound rotor, 225
written pole, 115, 116
motor–generator set, 117
motorized wheel, 246
MRI, 18, 155, 156
MTBF, 83

nanotubes, 62
40-mm long, 65
applications, 62, 64
carbon, 62
length, 64
multi-wall, 62
single wall, 62
NightStar flashlight, 168
NMR, 50
noise, 2, 81, 121, 135, 138, 139, 182, 183, 190, 218, 220, 222, 237

particle beam, 96
permanent magnet, 43
amount of material, 122, 124
annular, 163
embedded, 141, 142, 229
for medical devices, 138
NdFeB, 7, 43, 44, 47, 138, 144, 222, 229
rare-earth, 43, 44, 156
repulsion, 168
ring shaped, 160
rotating, 95
SmCo, 43, 44, 162, 163
stationary, 93
surface, 229
trapezoidal, 231
perovskite, 18, 50
phase diagram, 14
phonon, 17
photon, 17, 96
photovoltaic cell, 125, 248
Planck constant, 17
plaque excision, 147
portable
energy harvesting devices, 169
power generation, 162
power supply, 157
powder materials
Accucore, 36, 40
dielectromagnetics, 40, 111
magnetodielectrics, 40, 111
metallurgy, 41
soft magnetic, 36
Somaloy, 36, 43
power circuit, 91
power consumption, 183
power conversion, 164
power delivery, 213

power density, 71, 84, 105, 136, 160, 164, 172, 180, 183, 185, 207, 231, 246–248
power electronics, 2, 4, 67, 96, 130, 157, 235
power factor, 5, 109, 117
power interruptions, 117
power quality, 23, 117, 118
power range, 1, 157
power reactive, 185
power steering, 242
power system, 81, 89
power transmission, 155
prime mover, 95, 213, 217, 247
progress
   electrical machines, 1, 12
   material engineering, 1
propeller, 125, 189, 218, 221, 230, 249, 250
   counter-rotating, 219
   pitch, 225
   shaft, 220
propulsion
   electric, 213, 247
   for aircraft, 249
   for ship, 213, 215, 246
   marine, 173, 213, 218, 246
   mechanical, 213, 246
   railcar, 191
   shaft, 218
   submarine, 229
propulsor, 231
   CPP, 221
   duct, 232
   pod, 189, 191, 197, 214, 217, 220, 247
prostatectomy, 155
public life applications, 241
pump
   blood pump, 121, 140, 144
   classification, 141
   DeBakey, 142
   DuraHeart, 143
   electromechanical, 140
   hydraulic, 95
   implantable, 140, 156
   infusion, 135
   insulin, 135
   LNG, 79
   Streamliner, 142

submerged, 79

quantum mechanics, 17
Queen Elizabeth 2, 5, 216, 225
Queen Mary 2, 216, 221, 224
quench, 15, 61, 207

radar, 81, 97, 117, 202
recycling, 25
redundancy, 89, 218
refrigerant, 84, 87
reliability, 2
research programs, 2
residential applications, 238
retaining ring, 108
retaining sleeve, 81, 82, 111, 121, 162
robot, 122, 135, 153–156, 161
roller screws, 3
rotor bar, 104, 106, 108, 132, 192, 194
rotor end ring, 104, 105, 108, 192

Sayaka capsule, 150
Seiko kinetic watch, 167
self-powered microsystems, 164
semiconductor, 21
Shinkansen bullet train, 248
silicon, 21
   carbide, 21, 157
   content, 27
   laminated steels, 27, 82, 126, 144, 193, 253
   nitride, 157
small, 238
solar cell, 125
solar powered
   aircraft, 124, 126
   boat, 124
   vehicle, 124
solid rotor, 109, 113, 159, 206
solid state converter, 21, 198, 234, 246
superconductivity, 14, 15, 19
superconductor, 14, 19
   1G, 51, 53, 187, 190
   2G, 51, 53, 57, 209
   3G, 51
   amount, 184
   applications, 53
   BSCCO, 52, 53, 180, 207, 208
   bulk, 58, 197

HoBCO, 56
HTS, 17, 50, 53, 57, 58, 61, 171, 211
IBAD, 56
LTS, 17, 50, 51
MgB, 53
MgO, 57
monolith, 58
RABIT, 55
ring, 60
type I, 15
type II, 15
YBCO, 53, 55, 61, 191, 207
surface speed, 83, 104–106
surgery
  laparoscopic, 151
  minimally invasive, 151
  robotic, 152
switching
  capabilities, 19
  devices, 14
  frequency, 11
  speed, 19
synchropnous condenser, 211

thermal management, 97
thyristor, 19
tip driven fan (TDF), 237
torque
  accelerating, 194
  asynchronous, 115
  cogging, 7, 121, 128, 135, 190, 237
  density, 156, 184, 185
  electromagnetic, 72, 81, 118, 120, 184
  holding, 118
  hysteresis, 115, 128
  rated, 120, 191
  resultant, 120
  ripple, 124, 128
  starting, 194
  synchronous, 115

unmanned
  aerial vehicle, 250
  undersea vehicle, 231
  vehicles, 157

vector control, 11, 111
VentrAssist, 144

ventricle, 142
vibration, 2, 118, 130, 136, 156, 167, 220, 222, 235
video probe, 150

weapons, 162
winding
  armature, 7, 93, 190, 194, 206
  basket type, 121
  BSCCO, 189
  BSCCO field, 174
  cage, 108, 110, 111, 115, 131, 132, 192
  chorded, 92
  concentrated, 77, 92, 115, 124, 156
  concentric, 199
  copper, 163
  copper layer, 111, 113
  coreless, 190, 207
  damper, 93
  double layer, 92, 229
  dsitributed parameter, 92
  electroplated, 164
  field, 9, 93, 171
  fixture, 205
  HTS, 97, 193
  HTS field, 173, 183, 194, 198, 206–208, 218
  inductance, 121, 124, 162
  Litz wire, 162, 227
  LTS field, 203, 205
  non-overlapping, 77, 124, 156
  primary, 218
  racetrack, 185, 194, 198
  ring-shaped, 71
  saddle-type, 199
  salient pole, 77
  secondary, 218
  single-coil, 71
  skewed, 147
  slotless, 7, 121, 131, 187, 198
  slotted, 92
  stator, 75, 156
  suspension, 120
  three coil, 144
  three phase, 120
  toroidal, 143
  YBCO bulk field, 198
  YBCO field, 194